Disaster and Reconstruction

Originally published in 1982 and based on empirical research into the aftermath of the Friuli earthquake in Italy, the book reflects the perspective gained over a period of four years on the event itself and the subsequent response of the local population and national government. Unique insights were gained through one of the largest questionnaire surveys ever undertaken in a disaster situation and important questions are posed concerning the policies of reconstruction. Is a disaster 'the great equalizer' and does regional society emerge from it with redistributed power relationships, or are established structures reinforced? Who gets hurt and who benefits? What effects do poverty, regional remoteness from central government and the ethnic and cultural dimensions have on the situation? As a substantial treatment of a major catastrophe in all its aspects, this book will be of interest to students and researchers concerned with the impact of and response to natural hazards. It is based on a unique event, but the findings it reveals are relevant to all major catastrophes.

Disaster and Reconstruction

The Friuli (Italy) Earthquakes of 1976

Robert Geipel. Translated from the
German by Philip Wagner

Routledge
Taylor & Francis Group

First published in 1982 by George Allen & Unwin Ltd

This edition first published in 2024 by Routledge
4 Park Square, Milton Park, Abingdon, Oxon, OX14 4RN
and by Routledge
605 Third Avenue, New York, NY 10158.

Routledge is an imprint of the Taylor & Francis Group, an informa business

ISBN 13: 978-1-032-75191-7 (hbk)
ISBN 13: 978-1-003-47287-2 (ebk)
ISBN 13: 978-1-032-75464-2 (pbk)
Book DOI 10.4324/9781003472872

Disaster and Reconstruction

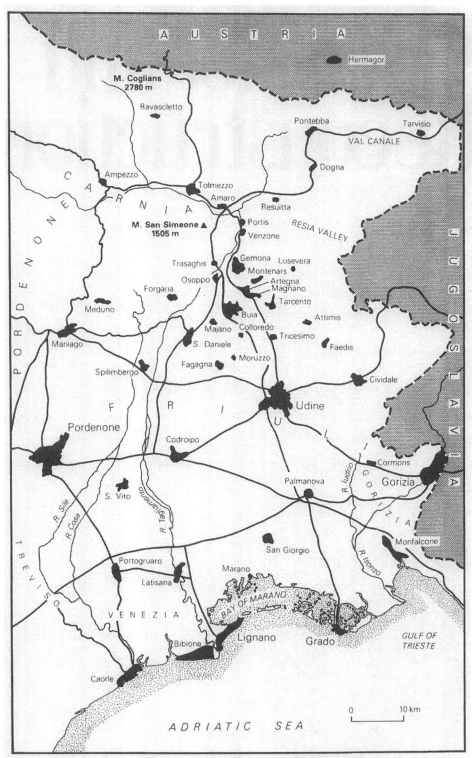

Frontispiece   General location map of the Friuli Region.

Disaster and Reconstruction

The Friuli (Italy) earthquakes of 1976

Robert Geipel

Geographical Institute,
Technical University of Munich

Translated from the German by
Philip Wagner,
Department of Geography,
Simon Fraser University, British Columbia

London
GEORGE ALLEN & UNWIN
Boston   Sydney

George Allen & Unwin (Publishers) Ltd,
40 Museum Street, London WC1A 1LU, UK

George Allen & Unwin (Publishers) Ltd,
Park Lane, Hemel Hempstead, Herts HP2 4TE, UK

Allen & Unwin Inc.,
9 Winchester Terrace, Winchester, Mass 01890, USA

George Allen & Unwin Australia Pty Ltd,
8 Napier Street, North Sydney, NSW 2060, Australia

First published in 1982

British Library Cataloguing in Publication Data

Geipel, Robert.
  Disaster and reconstruction.
1. Earthquakes–Friuli (Italy)—Social aspects
945′.39   HV555.I/
ISBN 0-04-904006-5
ISBN 0-04-904007-3 Pbk

Library of Congress Cataloging in Publication Data

Geipel, Robert.
  Disaster and reconstruction.
Includes bibliographical references.
1. Earthquakes—Italy—Friuli (Province) 2. Disaster
relief—Italy—Friuli (Province) 3. Disasters—
Psychological aspects—Case studies. 4. Anthropo-
geography—Italy—Friuli (Province)—Case studies.
I. Title.
HV600 1976   363.3′495   82-11625
ISBN 0-04-904006-5
ISBN 0-04-904007-3 (pbk.)

Set in 10 on 11 point Times by Preface Ltd, Salisbury, Wilts.
and printed in Great Britain by Biddles Ltd, Guildford, Surrey

To Gilbert White

*Who inspired and encouraged me
to go and look for myself
at what had happened in Friuli.*

Preface

The purpose of the present book on the earthquakes of 1976 in the Friuli area of northeastern Italy is to re-examine and extend existing knowledge about how a regional society behaves in a situation of catastrophe. It is indispensable in doing so to take careful account of the social, economic and political circumstances in Friuli as they were when the disaster struck, as well as at the time of reconstruction. The socio-economic consequences of the earthquakes, as they influenced people's perceptions and reactions concerning migration possibilities, the employment situation and altered residence and settlement patterns, will be discussed. This will then make it possible to attempt several broader generalizations about catastrophies, and to determine what has been learned in Friuli that would apply to other situations in a different context.

One of the important problems to be considered herein is whether or not it is possible for a catastrophe to provide an impetus to development for at least some sectors of a regional economy. Is a disaster 'the great equalizer', and does a regional society emerge from it with redistributed power relationships, or will established structures instead have become markedly reinforced? What groups are hurt most by a catastrophe, and which take advantage of it?

Such questions begin to arise when some time has elapsed between the usual hectic fieldwork phase and subsequent reflection and evaluation, particularly with subsequent studies. The original hypotheses then fall into proper perspective, and little-regarded details of the initial analysis can then be followed up. It is just as important to look for more general insights afterwards, when one is not burdened by details and can be detached from the immediate events, as it is to react quickly in a disaster situation, where unforeseeable developments are so hard to document and, under pressure to carry out rescue measures, nobody thinks of keeping track of just what is happening.

While the Friuli earthquakes were dramatic and tragic, they serve to illustrate the need for an understanding of natural disasters and behavioral responses to them. The past decade has witnessed a considerable increase in research interest in natural hazards in general and earthquakes in particular. In part, it is a reflection of the fact that there has been a number of very damaging events, accompanied by large loss of life or high losses of property and income. At the same time, certain shifts in social values have made it apparent that not only have past actions to deal with natural hazards been ineffective but also that the nature of the problem is becoming ever more complex.

This book is intended as a contribution to the quest for a better understanding of the nature of natural hazards and the way in which people respond to them. While it draws upon the findings of research undertaken previously, and especially in North America, it attempts to break new ground. Much of the previous work has been devoted to developing models of human reactions to disasters in a single city or parts of a city. In addition, many of the studies have not taken into account the problem of external influences, notably those which come into play in the wake of a sudden and massive disaster. The Friuli earthquakes affected nearly 100 urban and rural communities[1] spread over

4800 km^2. The impacts varied not only with physical circumstances but also with cultural, social and political characteristics. Moreover, because of the geographic location, adjacent to Austria on one hand and Yugoslavia on the other, it was an area where external political influence might be brought to bear, overtly or otherwise. At the same time, being geographically remote from the central government in Rome, the local inhabitants tended to be uncertain about the roles that should be or were played by people from afar in times of such disaster.

The results of the studies reported on here clearly indicate that the cultural context of natural disasters has an important and perhaps critical influence on how they are perceived and dealt with. It seems that, although there is considerable variation in physical circumstances in North America, there is a certain uniformity in perceptions of hazards and human responses to them. Thus, the perception of the flood hazard by householders in one city in the United States might resemble very closely those of residents in another city. Similarly, because adjustments relating to such disasters have often been institutionalized (in this case through the US Corps of Engineers), perceived solutions often have a certain uniformity. In the case of Europe, however, there is often considerable cultural diversity, even over very short distances. It is not surprising, therefore, that human responses to natural hazards differ sharply from one region to another and in many instances differ considerably from experiences on the other side of the Atlantic.

This book takes as a starting point some models which attempt to explain human reactions to natural hazards. Conceived by a number of researchers working in association with Gilbert White of the University of Colorado, they were derived from an examination of case histories in several parts of the world. To that extent, this book has the advantage of having potentially brought application to some of these models. At the same time, however, local circumstances, and especially those relating to cultural dimensions, may require a modification of particular aspects of these models. An important objective of the research undertaken on the Friuli disaster was to determine their applicability in a context where there was considerable cultural diversity, where external influences might play a large role, and where there was a high degree of poverty.

Models and methodologies generated by the North American group were applied to the Friuli case but, for reasons which are explained later, had to be modified considerably. An expanded formulation was developed by the German research team, drawing from (but adding dimensions which it was believed would help explain more adequately) the perceptions, actions and reactions in the European case under investigation.

ROBERT GEIPEL
Munich 1981

Note

1 Italy has four levels of public administration: the **nation** as a whole, **regions**, mostly consisting of a number of **provinces** (the region of Friuli–Venezia Giulia, for instance, consists of four provinces: Udine, Pordenone, Gorizia and Trieste) and **communes**. Communes mostly consist of various fractions or part-communes. Communes have a mayor, and provinces a prefect, who resides in a 'prefettura'. Regions are governed by a 'giunta' with a president as head. Self-administration is much less developed than in England or the USA.

Acknowledgements

This undertaking has been encouraged by Ian Burton's request for a completely new book for English-speaking readers, four years after the event, rather than simply a translation of the previously published German and Italian research reports,[1] so that a freer treatment of the material accumulated in those four years is possible. The author is grateful to numerous people in the catastrophe zone for the data secured, and to a large number of collaborators in Germany, and also colleagues in Canada and the United States, for making the preparation of the study possible.

Since catastrophies are in essence unpredictable events, it is hardly feasible to plan ahead a research design for a satisfactory scientific study. In the beginning, the author's trips to Friuli were like expeditions into largely unknown realms,[2] from which he brought back data that permitted certain conjectures which it would be too much to call 'hypotheses'. From one trip into the field to the next, contacts with official decision-makers multiplied, but also those with critical observers of what was happening, representing every social level. Official data collected were compared upon returning home to Munich with those previously gathered, and gave rise to new questions for subsequent visits to the area. It was of primary importance to establish a relationship of trust with local officials, the Emergency Commissioner from the central government in Rome and his collaborators. Among the latter, Consul Dr Facco Bonetti and Vice-Prefect Dr Toscano should be mentioned. But it was equally important to make contact with the residents of the area destroyed – in town halls and planning bureaus, in the building inspectors' offices, in schools for evacuated children, in the facilities provided for personnel engaged in clearing ruins and rebuilding, and finally in the prefab housing zones. It was necessary, furthermore, to establish and extend academic contact with colleagues associated with the affected areas by many years of fundamental study there, such as those at the Faculty of the Geographical Institute of the Facoltà di Lingue e Letterature straniere of the University of Trieste in Udine. Professor Giorgio Valussi. Director of the Department, with his penetrating studies of Friuli, contributed indispensable knowledge about the processes of mobility that were accentuated by the earthquakes of 1976. He provided the author with the opportunity for discussion with a large and interested audience in a conference at the University in Udine on March 2, 1977. He set up the organizational framework for the work taking place at the scene, arranged living quarters for the study group, and was always ready to provide instructive conversations or field excursions. He and Professor Guido Barbina[3] have themselves engaged in research on the earthquakes and the problem of the reconstruction. They facilitated contacts with groups concerned with commune policies as well. Signora Dr Giovanna Meneghel, Swiss-born, was the 'interpreter' in every sense of the word between the worlds of German and Italian geography, as well as an expert on the local area and a tireless organizer of teams of interpreters for the various survey operations. Dr Luciano Di Sopra and Dr Roberto Pirzio Biroli were particularly stimulating partners in conversations in their planning offices and in the field, as was Professor Raimondo Strassoldo in regard to matters of regional and ethnic sociology.

Our contact with the Friulian population at every level from decision-making officialdom to simple mountain peasantry was marked by spontaneous good will and helpfulness. The 'thank you' expressed to the student assistants and the research director by people whom we had hardly dared to interrupt in their tedious field labor, in the rebuilding of their houses, or in the drafting of site plans for temporary housing areas, who despite the language barrier gave us information, looked up statistics, or copied plans, was not for the research group alone. 'Grazie a lei' was meant primarily for the mountain engineer troops from Brannenburg and Passau, the Bavarian and German national Red Cross representatives, the ambulance personnel and field agents from Caritas, the Home Mission, the Worker's Welfare Society, and the World Lutheran League who were the real friends in their time of need. Their activities in Friuli had developed a human climate that often almost embarrassed the research group because of the sparse results they might sometime be able to publish, having enjoyed a degree of trust that they themselves had done nothing to create.

The progress reports from the early stages of the research facilitated useful contacts with the European Investment Bank in Luxembourg – which generously and unconditionally undertook the material support of the fieldwork before funding from DFG (Deutsche Forschungsgemeinschaft) occurred – and with international relief organizations interested in some check on the effectiveness of their aid measures. Such responsiveness intensified the motivation of the research director and his student assistants, but also made it clear that the results to be expected would affect some interested parties.

This brings us to help from German sources, only a few of which can be acknowledged: Inter Nationes as the source of funds for the translation, the Bavarian Red Cross for first contacts in the disaster area, the Bavarian State Office for data processing, colleagues at the Technical University of Munich, especially the graduate student group which actually constituted the research team in the disaster area; Richard Dobler, Falk Gottschalt, Rudolf Stagl, Michael Steuer. Helene Voelkl and Ursula Wagner. Material from their dissertations and MA theses has been used in many places in this book.

A word of thanks is due, finally, to colleagues in Canada and the United States. Just as the author, traveling in the Western United States in order to collect material on earthquake hazards, was on the point of visiting Boulder to look over what he had found and to discuss it all with Gilbert White,[4] the second Friulian earthquake of September 15 1976 occurred, and Professor White proposed that he should go and have a look for himself at what had happened. The 'first-hand look' soon took on the character of an obsession, for here was an appealing, friendly, hard-working people struck by a terrible emergency that a large number of helpful people from many countries were trying to cope with. Should not geographers especially figure among those doing what they could to help?

Ian Burton and Bob Kates, at an IGU working group conference on perception of the environment in Ibadan, Nigeria in August 1978, encouraged the author to rework the results already presented in his study and the further material from three subsequent field trips, for English-language readers. Ian Burton opened channels to George Allen & Unwin, and offered to undertake the editing. Philip Wagner, who had already participated in the fieldwork in February 1977 and had another look at Friuli in the summer of 1979, agreed to work on the translation, trying to preserve the spirit and ideas of the original, while rendering complex Teutonic sentence structures into not much less intricate English ones, pointing up, where possible, certain particulars of content of special interest. I am deeply grateful for his excellent work. These four English-speaking colleagues, and the many others whose influence is reflected only in the references, deserve especially warm thanks for their collaboration. Anne Buttimer, of Clark University, Nicholas Helburn, of the University of Colorado, David Ley of the University of British Columbia, and Derrick Sewell of the University of Victoria, BC, should be thanked in particular for their expert reading and criticism of the translation. The improvements are theirs, while the author accepts responsibility for remaining faults.

Notes

1 Geipel, R. 1977. Friaul – Sozialgeographische Aspeckte einer Erdbebenkatastrophe. *Münchener Geographische Hefte* **40**. Regensburg: Kallmünz.
   Geipel, R. 1979. *Friuli – aspetti sociogeografici di una catastrofe sismica*. Milan: Franco Angeli Editore.

Geipel, R. (ed.) 1980. *Il progetto Friuli – Das Friaul Projekt 'ricostruire' – quaderni di echistica*. Udine: Martin Internazionale.

Steuer, M. 1979. Wahrnehmung und Bewertung von Naturrisiken am Beispeil zweier ausgewählter Gemeindefraktionen im Friaul. *Münchener Geographische Hefte* **43**. Regensburg: Kallmünz.

Dobler, R. 1980. Regionale Entwicklungschancen nach einer Katastrophe. Ein Beitrag zur Regionalplanung des Friaul. *Münchener Geographische Hefte* **45**. Regensburg: Kallmünz.

2 The fieldwork in Friuli received considerable support from the Bavarian Ministry of State and the Bavarian Red Cross, whose intercession made the first contacts possible.

3 Barbina, G. 1976. Teoria e prassi della ricostruzione. *La Panarie* no. 33–4, 23–31.

Valussi, G. 1977. Il Friuli di fronte alla ricostruzione. *Riv. Geog. Ital.* LXXXIV, 113–28.

4 Generous support from the DFG made it possible for the author to work with the institute mentioned. He would like to express his thanks to Kenneth Boulding, Eugene Haas and Gilbert White for extensive discussion and access to unpublished manuscripts and working papers. The ideas expressed herein are to a large extent derived from that exchange of views In the meantime, most of these ideas have been published in:

Haas, J. E., R. W. Kates and M. J. Bowden (eds) 1977. *Reconstruction following disaster*. Cambridge. Mass.: MIT Press.

Friesema, H. P., J. Caporoso, G. Goldstein, R. Lineberry and R. McCleary 1979. *Aftermath. Communities after natural disasters*. Beverly Hills: Sage.

Wright, J. D., P. H. Rossi, S. Wright and E. Weber-Burdin 1979. *After the clean-up. Long range effects of natural disasters*. Beverly Hills: Sage.

Contents

List of tables

1 The disaster area and its people

1.1 The earthquake events

On May 6, 1976, at 9 p.m., an earthquake with a strength of 6.4 on the Richter scale and lasting for almost a minute took place in the area of Friuli in northern Italy. Its epicenter lay only about 5 km below the surface (Fig. 1.1). The number of dead and wounded was very much affected by the particular time of day and the weather: most people were at home but out of doors because of the heat of the early summer evening. Nevertheless, 939 people perished in the wreckage of the 17 000 collapsing houses, and 2400 were injured. The homes of 32 000 people were totally destroyed (Fig. 1.2). Several hundred tremors, aftershocks and heavy rain that set in immediately after the quake pulverized the already badly damaged buildings still further, making the homes of another 157 000 people uninhabitable. Altogether, an area of 4800 km², embracing nearly 100 communes with a population of half a million, was affected by the catastrophe, and a zone 25 km across, containing 1766 km², was totally leveled. The first, somewhat exaggerated, estimates placed the damage at 4400 billion lire, or $6 billion (i.e. American billion).

The initial relief measures were favored by the fact that around two-thirds of the Italian army were holding maneuvers in northeastern Italy at the time of the catastrophe, and NATO troops and Austrian and German relief organizations arrived quickly on the scene. Camps with a total of 16 000 tents sprang up, railway cars were brought into the ruined settlements and small mobile campers were set up. An emergency commissioner sent from Rome, State Secretary G. Zamberletti, assumed charge of the administration in Udine and Pordenone provinces and set up emergency headquarters in the provincial capital of Udine.

There was no mass flight from Friuli despite the extent of the damage, even though after centuries of emigration many Friulians living all over central and western Europe, the United States, Canada, Australia and South America keep close connections with the homeland, and on this occasion offered their hospitality and suggested their kin leave Friuli and join them. On the contrary, the people wished to begin reconstruction forthwith and to bypass the phase of merely temporary accommodation.

The fears of Friulians that the fate of victims of Sicily's Val Belice earthquake of 1968 might be repeated in their case gave rise to the motto 'Dalle tende alle case' (*straight* from the tents into the houses – and not into barracks first).

An end came abruptly at 11:30 a.m. on September 15, 1976 to the four and a half months of continuous and energetic rebuilding of houses and especially factories (Fig. 1.3). After numerous light shocks, a second earthquake, at a strength of 6.1 Richter, struck Friuli again. The number of homeless, which had decreased by the beginning of September to 45 000 because of the repair of less badly damaged buildings, and because people found shelter with friends and relatives elsewhere, swelled again to more than 70 000. Even though the quake was weaker than the first on May 6, the psychological effects on the people's spirit of resistance were much worse this time. Bad

1

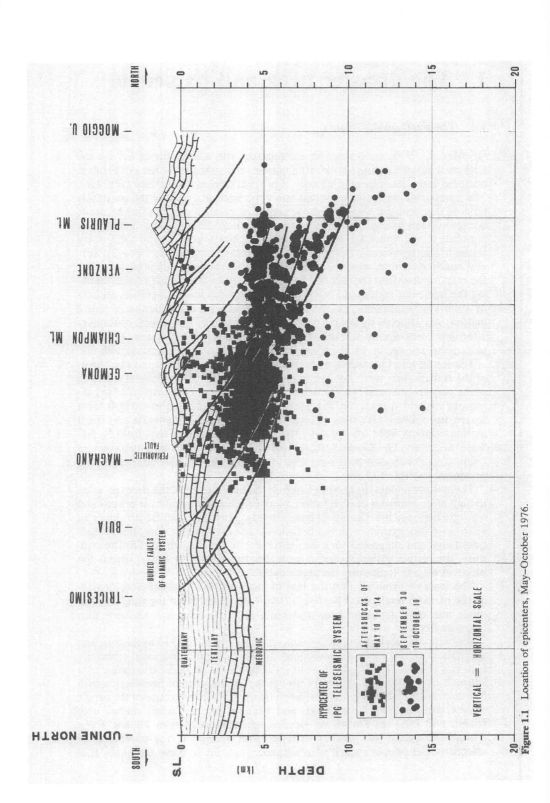

Figure 1.1 Location of epicenters, May–October 1976.

Figure 1.2 The totally destroyed city of Gemona, with a former population of 12 000 inhabitants, lies within a frame of high alpine mountains and its first outliers of mediterranean vegetation show transition to the hills of the Tagliamento moraine. On the alluvial fan, the soil of which had not been totally consolidated, even modern buildings collapsed. With about 300 dead, Gemona suffered about one-third of all casualties in Friuli. A city full of medieval monuments was forever lost as a center for sightseeing and tourism.

Figure 1.3 It is part of Friuli's tragedy that, after the endeavours of the population to rebuild their houses after May 6, the second earthquake of September 15 frustrated all initiatives. So many attempts to reconstruct had to be given up. Ruins with makeshift repairs show that the inhabitants cherished hopes until the last minute that they would succeed in rebuilding their picturesque farm houses before winter.

landslides blocking escape routes out of the mountains, the loss of savings that had already been invested in reconstruction, the dangers of the prospect of the hard mountain winter, all overwhelmed the resistance capacity of a mountain population accustomed to privation. Although the loss of life this time was much less (only 11 additional fatalities), and despite the fact that nearly 400 tremors between the May and September earthquakes had made the inhabitants aware of the riskiness of the ruined buildings, it was only at this point that the great exodus from the afflicted area began.

The Emergency Commissioner installed on May 7, who had planned relief measures only until July 27, was nevertheless re-appointed on September 13 because the regional administration was totally swamped, and he received expanded powers once more after September 15. He was thus able to requisition hotels and apartments in the Adriatic coastal towns, occupied only during summer and standing empty in winter, as housing for the elderly, the disabled, and women and children. Many able-bodied people were also evacuated to the coast, however, and had to put up with long daily journeys to their jobs within the disaster area. Farmers who did not wish to leave livestock unattended, as well as persons who were determined to get on with the rebuilding of work places and homes, mostly refused evacuation, however, and spent the unusually cold winter of 1976–7 in primitive emergency quarters. Illness due to the cold and to overwork was common, not only within the disaster area but also in the coastal towns – foggy, chilly and inhospitable in winter – where many of the hotels simply had no heating.

Table 1.1 and Figure 1.4 clarify the course of the evacuation process and point up some of the 'benchmarks' which we used in an attempt to forecast the progress of the whole evacuation-and-return operation. Figure 1.5 shows areas of origin of evacuees in coastal centers.

Lignano and Grado bore the brunt of the evacuation. Since people in the northern mountainous communes refused to be removed to an altogether

Table 1.1 Evacuation sequence.

Day	Lignano	Grado	Bibione	Jesolo	Caorle	Ravascletto	Total
9.16.76	5 267	1081	192	150			6 690
9.20.76	11 811	2289	3695	745	108		18 648
9.25.76	17 120	4108	4431	1160	400		27 219
9.30.76	18 125	5068	4620	1278	499	302	29 892
10.15.76	19 300	6226	4116	1411	480	722	32 255
10.31.76	19 252	6480	3687	1376	443	808	32 048
11.15.76	18 285	6449	3318	1326	359	826	30 563
11.30.76	17 149	6290	3026	1183	324	837	28 809
12.15.76	15 772	6065	2657	1094	276	835	26 699
12.31.76	15 156	5681	2459	1031	255	801	25 383
1.15.77	14 175	5365	2271	862	243	759	23 675
1.22.77	13 730	5196	2118	805	230	751	22 830
2. 7.77	12 751	4762	1838	655	225	665	20 896
2.21.77	11 109	4310	1375	476	176	580	18 026
2.28.77	10 031	3839	1204	337	146	428	15 985
3. 7.77	8 710	3454	983	266	85	397	13 895
3.14.77	7 462	2899	836	203	77	386	11 845
3.21.77	6 209	2562	678	170	61	316	9 996
3.26.77	5 013	2280	519	145	52	274	8 283
4.17.77	935	261	106	78	27	26	1 433

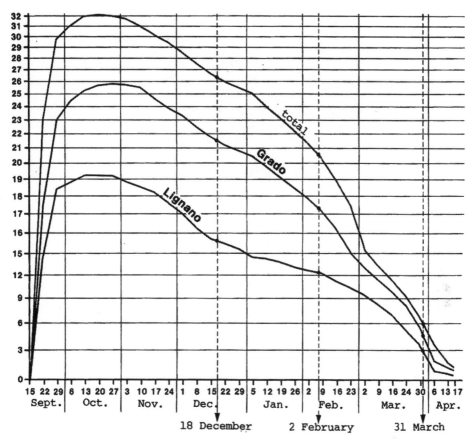

Figure 1.4 The sequence of the evacuation operation.

unfamiliar environment nearly 100 km away, hotels in the mountain winter sports resort of Ravascletto were also commandeered for 800 homeless.

Reconstruction plans after the first earthquake were hampered by the Italian parliamentary elections set for June 20, discouraging politicians from carrying out unpopular measures such as identifying and requisitioning areas suitable for building barrack settlements, or from deciding not to plan for any further reconstruction in places that were too remotely located. Decisions could not be put off any longer after the second earthquake in September, however, and the seismic security of places chosen for the rebuilding and the setting up of temporary settlements was, by then, a factor to be considered. At the end of 1976, the situation was as follows:

(a) 25 000 people were already living in newly erected prefabs, temporary shelters and railway cars;
(b) 15 000 were occupying small camping trailers;
(c) 1000 still dwelt in tents;
(d) 25 000 remained in the evacuation centers on the Adriatic coast.

Thus, housing space had to be created for some 66 000 people. A major prefabricated housing program made provision for about 21 000 units, of which 49 per cent were to come from the Emergency Commissioner, 45 per

5

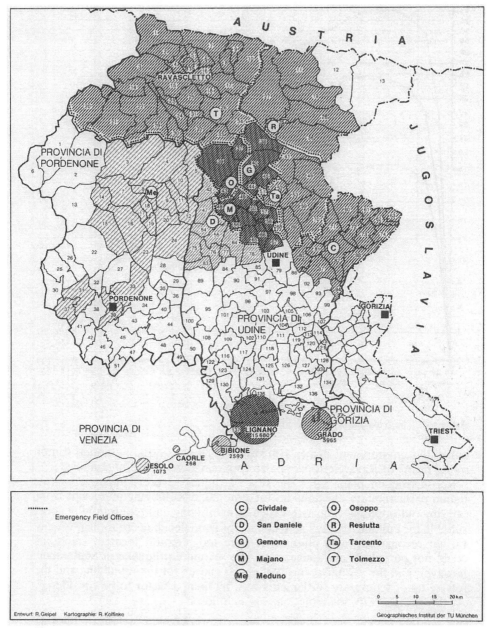

Figure 1.5 Areas of origin of evacuees in coastal centers.

cent from the regional administration, and 6 per cent from mostly foreign governmental and private relief organizations (Fig. 1.6).

Despite the hindrance of an extremely severe winter, the evacuees were successfully returned from the coastal towns by mid-April, as Figure 1.4 shows. Barely enough time had been left to get the apartments and hotel

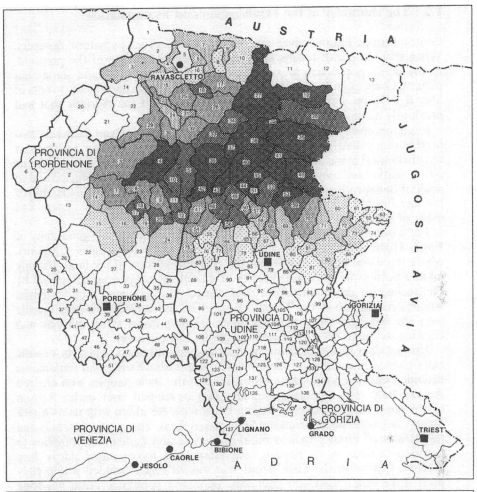

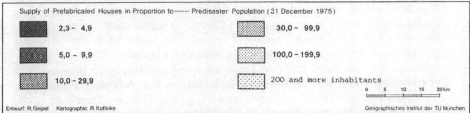

Figure 1.6 Supply of prefabricated houses.

rooms that were thus released ready for the vacation season. The regional administration could certainly not afford to disregard the income from tourism, which was one of the region's largest sources of foreign exchange. Whatever shortcomings of the barrack towns may have become apparent later on, the return of the evacuees at the date that had been planned was a major logistical achievement in coping with a catastrophe on such a scale.

1.2 The character of the Friuli region and its population

Central Europe has the advantage of being rarely struck by natural disasters. When they do occur, results are especially severe because of the predominantly dense population and the age of the building stock and settlement pattern. Consequently, the attention of the world was focused by the events of May 6 1976 on a small zone at the northern end of the Adriatic that had previously made few headlines.[1]

Friuli occupies the northeastern corner of Italy, wedged between two neighboring countries of distinctive political characters: Austria, which must remain neutral towards all alliances because of its peace treaty and despite its purely market economy, and Yugoslavia, also able to remain unaligned and to assert its independence of the Eastern bloc because of the skillful leadership of the late President Tito, even though it has a socialist economic system and many of the features of a communist society.

The eastern Alps indirectly and directly dominate the whole geography of Friuli. High, rugged Alpine chains that converge on the Triglav (2863 m), where three countries meet, frame the northern portion of the land. Its hilly middle portion in the main consists of glacial till heaped up or spread by Pleistocene ice descending from those ranges, and is watered by torrential streams converging in the Tagliamento, which has spread a boulder mantle over the heart of southern Friuli after crashing down steep canyons and chewing its rough course through the glacial amphitheater.

This area has had a special character since ancient times, when its Venetic culture stood in contrast to that of the Etruscans whose trade and settlements extended into the north-east, and to those of the Italic peoples who created Roman Italy. The Romance language that developed later under Roman rule is not a simple dialect of Italian but a separate idiom with its own rich literary heritage and vigorous modern spoken form. The mountains, hills and floodplains of Friuli provided military recruits and agricultural supplies to maritime Venice in its heyday, and later were incorporated partly into Austria. Only after the First World War was all Friuli included within Italy itself. Like other mountain countries, poor and populous, Friuli has long depended on its mercenary soldiers and itinerant craftsmen in the building trades and others to seek remittances abroad to supplement its own meager resources but, as ensuing pages will reveal, the 'sense of home' is strong among Friulians worldwide, and many wanderers return.

Whereas to the north and east Friuli has comparatively open frontiers with neighboring countries, its southern boundary is the Adriatic coast, where some of the tourist centers such as Grado, Lignano, Bibione, Caorle or Jesolo may figure on the mental maps of anglophones. Trieste – the very unloved capital of the autonomous region of Friuli–Venezia Giulia, which had its Golden Age as an Austrian imperial port city but declined after the First World War and was, for a good while, disputed between Italy and Yugoslavia after the Second – is even better known. In a 'marriage of convenience' with Friuli, Trieste, which is the largest city of the area and its capital although peripheral to it, looks southward and has little to do with its hinterland, the real center of which is Udine, a lively town of over 100 000 inhabitants. The latter is also the administrative center of the most populous (300 000) of the four provinces that make up Friuli–Venezia Giulia. (Pordenone Province has *c.* 260 000; Gorizia 150 000; and Trieste just under 300 000.) Among the heterogeneous population of this region of 1 200 000 people, about 400 000 are ethnic Friulians, speakers of one of the distinctive Romance languages preserved only here and there in the Alpine area, including the Dolomites (Ladin, 30 000) and the Swiss Cantons of Engadin and Graubünden (Romansch, 43 000).

Friuli consists in approximately equal measure of high mountain zones in the north and hilly lands in the center, as well as plains in the south. The Friulian ethnic core lies in the mountains and the hill country. Venetian influence prevails in the south, along with the homogenizing transformations brought about by tourist centers, so that *Furlans* is hardly spoken there any longer. On the Yugoslav frontier there are Slovenian and, in the north German-speaking, minorities.

The earthquake affected almost exclusively the portion of the region inhabited by the Friulians and laid to ruin their rich cultural heritage.

It was the middle mountain country and the hills that the earthquakes shook and shattered, turning centuries-old cities filled with splendid monuments and a whole artistic heritage to a dusty rubble, for there along the line of hills the belt of medieval and Renaissance cities mostly lay. The shocks were lesser in the high Alps, in the borderlands of Yugoslavia, and in the lower Tagliamento Valley and the well watered but only recently reclaimed coastal plain and delta region.

The Alps condition climate, too, for this area of north-east Italy, with up to

Figure 1.7 The River Tagliamento cutting through the Carnian Mountains, which rise up to 2000 m, is a torrent with very irregular discharge meandering through its shoals of gravel. The small, wooded hill at the right of the gorge is crowned by the fortress of Osoppo. Gemona sits on a triangular alluvial fan. In the background of the gorge, the rockfall slopes rise above Venzone. This transition zone between mountains and hills was the core of the disaster area.

9

3000 mm annual precipitation, is among the rainiest parts of Europe and gives the lie to the notion of 'la bella Italia' in this respect. This precipitation, especially in the mountains, for much of the time falls as snow. (Many of the organizations that sent prefab huts tended to overlook this important fact.)

The major river of Friuli is the Tagliamento (Fig. 1.7), which so fascinated Ernest Hemingway that it became almost a major character in two of his best known novels, *A farewell to arms* (1929) and *Across the river and into the trees* (1950). Both of these war novels, so characteristic of their author, give an idea of what a fought-over country Friuli has been, not just in the tragedies of the two great World Wars in Europe, but throughout a long and often cruel history as both a gateway for peoples out of the east and a contact zone of Latin, Slav and German. The ruins of the great Roman city of Aquileia attest to a landscape of conflict, as do the splendid fortifications of Palmanova, by means of which the Doges of Venice tried to secure their empire of the sea against land powers, and which became the model for the fortified bastions all over Europe in the age of the Baroque.

A troubled history and peripheral position within Italy have meant protracted periods of economic stagnation and compelled many Friulians to emigrate. Just as a modest degree of prosperity was beginning to become apparent in the form of numerous newly established enterprises, mostly producing consumer goods, and Friuli began rather belatedly to participate in the increasing well-being of northern Italy, the earthquake struck on May 6 1976 and wiped out whatever had been attained thus far.

This destruction all took place in a region where the capacity to recover appeared to be limited by several factors. Within the overall European framework, Friuli is the easternmost outpost of the European Community, comparable to the Bavarian Forest (Bayerischer Wald) in its exposed position, situated at the junction of three boundaries and adjoining a country which is not a member of the European Community (Austria), as well as one with a different economic system (Yugoslavia). Certain political tensions, e.g. the South Tyrol problem and the question of Trieste, existed at times among these countries. The influence of the highly developed area of the North Italian Plain indeed stretches out along the Alpine border city axis of Torino–Milan–Brescia–Verona–Vicenza–Padua–Venice–Trieste toward the eastern part of northern Italy, but with declining strength. Only Pordenone and Udine, and to some extent Gorizia, belong to the development axis where there is a high degree of industrialization. The disaster area lies north of this development axis and is connected with it by a heavy commuter movement (Fig. 1.8). Even long before the earthquake catastrophe, the mountain and hill districts had undergone a partial loss of population by migration over the short distance to this development axis, and this was beginning to create a strain on the infrastructure there, e.g. on school and hospital facilities, while the area north of the axis was marked by a 'retirement home effect' and the seasonal or lasting absence of the younger elements of the labor force.

The other Friulian development zone is the Adriatic coast, featuring a massive building boom and a three- or four-month tourist season, with about 150 000 hotel beds and apartment spaces, mostly belonging to condominium companies with many of their shareholders living in Germany, Austria, Switzerland, or elsewhere in northern Italy. There are more jobs here again, although they are only seasonal.

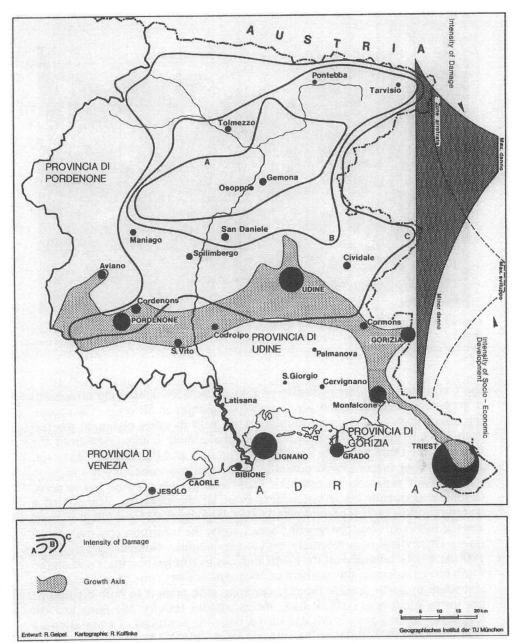

Figure 1.8 Disaster area and growth axis (courtesy of L. Di Sopra).

If we were to construct a regional income profile on a 100-km straight line from the Austrian border to the Adriatic, we would discover a stable peak on the curve around the Udine–Pordenone complex and a seasonal one for the coast, while the northern part of Friuli would fall far below in income. That area's tourist potential has to stand up to the competition of the mountain resorts of the Dolomites immediately to the west, and the water sport area of

11

Figure 1.9 After the inhabitants have left their destroyed cities, and while the bulldozers have stopped working on a Sunday, a macabre earthquake tourism develops, drawing sightseers from faraway places.

the Carinthian lakes on its northern side, as well as sightseeing attractions nearby, such as the Adelsberger Caves near Postojna in Slovenia. The charming hill country of the glacial amphitheater, with its cities of tourist interest such as Gemona and Venzone, faced overwhelming competition from the south, where there are cities like Venice, Aquileia and Palmanova. However, Friuli's former urban tourist potential is now entirely gone.

As already pointed out, northern Friuli as a peripheral area has to serve destinations outside itself but suffers from all the negative elements of a transit corridor, such as overcrowded roads, highway rest stops, filling stations and the billboards associated with them. Lately, the transition area has turned into a Disneyland of involuntary prefab expositions.[2] Comparing Friuli to a 100-km square bounded on the north and east by foreign countries and on the south by the Adriatic, the gradient of economic activity runs from west to east and south to north, so that the disaster area, with respect to both directions, lies in the quadrants farthest from the maximum activity. We have to look sceptically at the region's attractions, for over the centuries of emigration its ability to hold its population has proved weak (Fig. 1.11).

This same peripheral position in the extreme north-east of Italy had a rather discouraging effect on investment, and handicapped Friuli in the struggles with more centrally situated areas for economic growth. The peninsula of Italy has problemless land borders with its neighbors in the cases of friendly France and neutral Switzerland and Austria, whereas Yugoslavia represents a point of potential crisis in view of Soviet incursions into Hungary, Czechoslovakia and Afghanistan. So the Italian state concentrates a great part of its defense effort on this nervous frontier. Military bases, airfields, defensive installations and maneuver areas have produced restrictions on building and

12

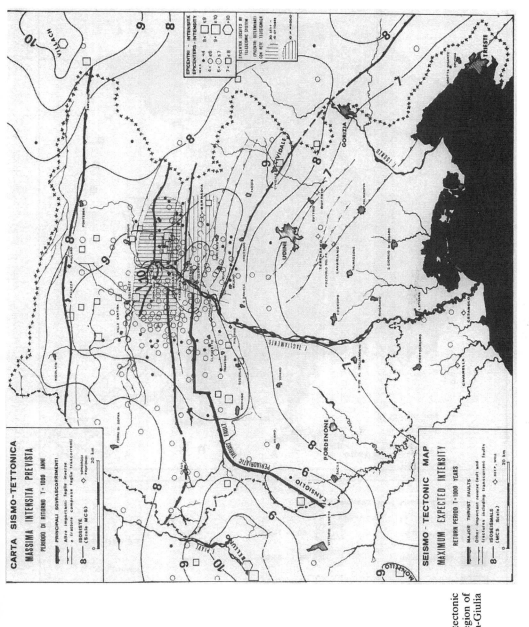

Figure 1.10 Seismo-tectonic map of the Region of Friuli–Venezia-Giulia

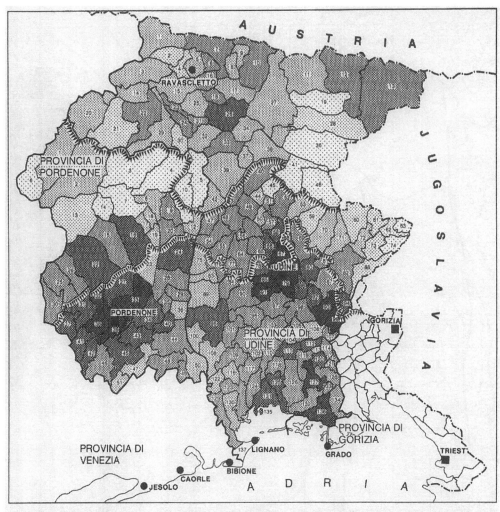

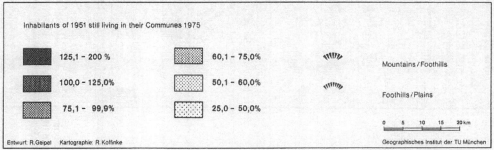

Figure 1.11 Change in population between 1951 and 1975.

regulation of land use. The presence of so many young military personnel from all over Italy poses a threat of conflict with local people, and makes the ethnic Friulians draw closer together in the face of the outsiders.

The difficult economic situation and high birth rates typical of a conservative, very rural and devoutly Catholic population have encouraged emigration

14

Table 1.2 Periods of emigration.[5]

Periods	Numbers emigrating
first period (1876–80)	88 466
second period (1881–1915)	1 319 327
third period (1916–40)	311 356

waves ever since the middle of the last century. Between 1876 and 1940, nearly 1.8 million Friulians left home,[3] permanently or temporarily, to begin a new existence somewhere else in Italy or more commonly abroad (Table 1.2). At first Switzerland, Austria, France and Germany were the most popular destinations, but later Argentina, Brazil, Canada and the United States were most prominent.[4]

After the Second World War, the strong economic growth in Lombardy and also abroad in German-speaking lands kept the emigration process operating. The percentage of persons employed in agriculture in the Friuli–Venezia Giulia Region declined from 26.1 per cent to 12.4 per cent between 1959 and 1971.

It is customary in Friuli for an able-bodied man to earn enough money abroad while he can to maintain his family at home, and to make provision for his old age by either building or keeping up a house. The sense of community is very powerful even when abroad, and is accentuated by employing a distinct Romance tongue. Its approximately 400 000 members had established a strong awareness of the region's unity even before the regional statute promulgated in 1970 for this and other regions of Italy. Their status as a frontier folk between three countries colors the historical experience of the Friulians and unifies them. Their habit of settling close to each other abroad makes for the establishment of receptive zones in various parts of the world, that could provide the contact points for a possible accelerated emigration following the earthquakes. These islands of settlement keep telephone contact among themselves and with Friuli.[6]

A border population with rather traditional and conservative ways, particularly among the more isolated mountain folk, in many countries may provide fertile ground for antipathy toward the central regime. Such is the case of Scotland with regard to 'London', in Bavaria with respect to 'Bonn', and in Friuli regarding 'Rome'. The earthquake catastrophe stimulated this emotional feeling of distinctiveness even more. There is a separate regional political party that tries to channel and cultivate this sentiment. The Commissario Straordinario became the symbol of Roman centralism, pitted in this situation against a heightened regional autonomy. The political situation prevailing in Italy at the time, furthermore, gave rise to an unease in conservative circles over the 'historic compromise' between the Christian Democrats and the communists that was then unfolding. Such circumstances can give an impetus to a stronger emigration tendency. According to the Emergency Commissioner, 20 000 people disappeared from Friuli for a while after the earthquake, some going, for example, to stay with relatives in Italy and abroad, sometimes on the lookout for better living opportunities outside the afflicted district. On the other hand, many people working abroad came back to the earthquake area for the time being, so that there were particularly active outside contacts during this period.

From a research standpoint (cf. the questionnaire and results reported on p. 83), we were particularly interested in whether the earthquake disaster would further strengthen this latent disposition to emigrate, or if the modest trend toward returning to the homeland and the new job opportunities created there, apparent not long before the quakes, would persist.

1.3 Problems created for planners and decision-makers

The planner approaches the subject of a disaster with a particular ambivalence. On the one hand, it represents to him the absolute negation of his previous professional activities: residential areas are leveled, jobs lost, support facilities destroyed, traffic routes blocked, and hoped-for growth is turned instead into emigration or even mass flight. The rarity of such an event makes it clear that no routines can be relied upon by the planner to deal with the disaster.[7]

This rarity of the event, on the other hand, provides the opportunity for putting forth proposals for original solutions to mitigate the results of the disaster, and 'after the cards have been more or less reshuffled', to be able to plan and proceed from a new starting point, more freely than under the former impediments. Thus, disasters can be seen as notably rare events that bring about large changes in the development process. The dismantling of many German industrial enterprises, with the later reconstruction with better mechanical equipment that this demanded, can be seen in retrospect as a decisive stimulus to economic growth in the Bundesrepublik after the Second World War, even though dismantlement at first represented a disaster for those who lost their jobs by it. The same applies to Germany's war-ruined cities and their rebuilding during the period of the 'economic miracle'.

Rebuilding plans after a disaster pose problems for the affected communities. Some of them have already been critically considered in the book by Haas et al. (1977) on *Reconstruction following disaster*; other books such as *Aftermath* or *After the clean-up* have looked into the long-range effects of natural disasters (see Acknowledgements, note 4). In the book by Haas et al., for example, the following problem areas stood out:

(a) People in whose immediate families fatalities occurred are interested in avoiding a repetition of the catastrophe and demand a better-conceived, safer city.
(b) This conflicts with the desire of the less affected to go back as soon as possible to normal conditions of 'orderliness'.
(c) Similar attitudes are found among evacuees willing to return, and some community leaders, who see people living in emergency housing as a standing reproach.
(d) These demands go against the wish of the planners to make careful choices amongst alternatives.
(e) Ranged against the planners are less heavily affected investors who want to extract a competitive advantage over heavily hurt rivals through fast deals.

Major issues are thus involved right away. For example, should rebuilding be organized by an outside crisis staff from higher up (in Friuli, this meant

Commissario Straordinario Zamberletti from Rome) or by the existing decision-makers from the affected provinces? In order to generate income in the stricken area after the catastrophe, local firms, in so far as they are still in a condition to operate, should be given preference in the reconstruction. In Friuli, instead, there was a massive importation of prefabricated houses from other areas and even other countries and continents (Canada).

The modifications of the decision process called for in this particular situation took on special importance in Friuli and deserve closer consideration.

A crisis staff legally installed, with its headquarters in Udine[8] in the disaster zone, by the central government in Rome, invoking its constitutional powers, was given precedence over the regional authorities of the autonomous Friuli–Venezia Giulia region and the prefectures of both of the affected provinces of Udine and Pordenone. The role of this public servant in the emergency brought to a head all the problems and tensions between political centralism and aspirations toward local autonomy in Italy. Embittered by the feeling of being shoved aside and plagued with a host of problems associated with the catastrophe, Friulians, on occasions, revealed by poster campaigns their determination to protect their regional autonomy. Among the posters some, to the astonishment of neighboring Austria, preoccupied with problems of the South Tyrol and the Slovene minority in Carinthia, proclaimed affection for the Austro-Hungarian monarchy, 60 years after its dissolution.

In view of the experience of the Friulians in partisan campaigns in the Second World War, proclamations such as 'We want a German government' can be taken not only as spontaneous expressions of gratitude for effective help from West Germany, but also as criticism of the officialdom of their own regime. The attitude toward the Emergency Commissioner altered in time from initial coolness to ovations when he left, and the regional and provincial authorities who had been deprived of power went from perfunctory greetings at first to envy of his competence and jealous criticism of the effectiveness of his decisions.[9]

In contrast to such criticism, to outside observers the effectiveness of catastrophe management seemed astonishingly high. Accusations of corruption made against some colleagues and mayors can do nothing to change this. An elite of civil servants from many different parts of Italy was brought together and established both in the center of organization at Udine and in command posts at Cividale, Gemona, Majano, Meduno, Osoppo, Resiutta, San Daniele, Tarcento and Tolmezzo and, in addition, in evacuation sites at Bibione, Caorle, Grado, Jesolo and Lignano on the coast and Ravascletto in the mountains. This staff, whose members often lived separated from their families for months in the disaster area, developed a strong group spirit, and sometimes had tense relations with local administrations. The surprisingly decisive measure of compulsory acquisition of substitute housing space in the hotel cities on the Adriatic coast could, in particular, hardly have been carried through by regional authorities. The guaranteed release of the space by March 31, 1977, at the start of the tourist season, was somehow a compromise between disaster aid and the wish for tourist profits, but it was at the same time the means of gaining a little breathing space for coming to certain decisions, because when there is a definite time set the decision-makers are under pressure to meet their goals.

The situation following a disaster may be represented by a supply-and-

Table 1.3 Synopsis of demand and supply for aid measures.[9]

Public demand	State aid supplied
0–1 hour	9 p.m., May 6, 1976.
within 24 hours	Rescue of people from collapsed buildings; life saving; fire protection measures; safety test of major bridges etc.; provision of tents, camp beds and blankets.
48 hours	Medical aid, operations, amputations; water supply; care of aged, children and the ill; issuing of emergency decrees.
3 days	Burial of the dead; removal of animal carcasses to prevent epidemics; sanitary installations; identification of unsafe structures.
within 1 week	Emergency housing for the able-bodied population; distribution of aid supplies; closing off of ruined buildings; public security against looting; protection of artistic treasures.
	State and foreign aid
within 1 month	Clearing of rubble under reconstruction plans; systematic application of foreign relief supplies; major administrative regulations and regularized law-making.
3 months	Selection of areas for rebuilding; restoration of infrastructure (schools, hospitals etc.); foundations laid for prefab cities.
0–2 hours	10.21 a.m., September 15, 1976: second earthquake.
24 hours⎫ 48 hours⎭	Mass evacuation to the Adriatic coast.

demand model for aid measures undertaken by public officials and agencies, in which the affected population generates demand at certain times and over particular periods for given kinds of goods and services, and the government is evaluated critically according to its ability to deliver these services. The example of the Friulian earthquake took the form shown in Table 1.3.

If these services are delivered only after these deadlines, their usefulness is lost. If they are altogether not forthcoming, the natural catastrophe turns even more into a social one and trust in the regime is lost. On the other hand, if these services are furnished by someone other than the 'constituted authorities', for example helpers from abroad, then people may feel more loyal to the latter than to their own administration, whose measures have proved too weak (Fig. 1.12).

For this very reason, totalitarian regimes prefer to forego all external aid rather than give their people the chance to enter into contact with outsiders, especially those who come to help and do not fit clichés of friend and foe.

Foreign disaster aid in Friuli, although free from any such problems and an example of European solidarity which even took in the Eastern bloc countries, was nonetheless not lacking in well intentioned mistakes. It is possible to avoid these by flying experienced civil servants from abroad into the disaster area immediately, so they can report on the kinds and amounts of relief supplies needed. If this cannot be done, the request of authorities within the stricken area should be awaited, so that really needed help can be given. This calls for a particular brand of logistics, especially in the case of a disaster on the scale of Friuli.

The process of coming to terms with the catastrophe there was made harder by the fact that the second earthquake on September 15 1976, 132 days (4½ months) later, set off a second wave of events and reactive measures leading to interference effects between the two respective time schedules. These

Figure 1.12 Osoppo, with a population of 2500, was nearly totally destroyed. Its inhabitants show their gratitude to rescue teams from the Federal Republic of Germany, to American and Canadian NATO personnel, and to helpers from France.

interference phenomena again made the installation of the Emergency Commissioner necessary, on the basis of Article 5 of Law 996 of December 8, 1970. This took place on September 18. His period of authority ran until April 30, 1977 and was valid in all districts named in Decree 0714.[10]

The changes and tightening up in legal powers of decision-making affected all other issues negatively, particularly the changes in land use.

Changes in land use, extending to the reconsideration of the future viability of whole settlements, or at least portions thereof, were absolutely necessary in Friuli. The settlement pattern there is based on small towns and large villages in the plains, the hilly zone and the major mountain valleys. There are hillside villages, hamlets and isolated groups of farmsteads in the mountains. This settlement pattern arose out of a milennial process of natural resource exploitation mostly centering on agricultural land-use. Only the traffic functions of the communities in major mountain valleys, and the gradual process of mostly small and microscale industrialization had to some extent begun to create the occupancy pattern of an industrial society in the area, during the period just preceding the catastrophe.[11]

The desire of the population to remain, and the necessity the Special Commissioner had to face of arriving at his decisions in the interval between the onset of winter and the beginning of the tourist season in the coastal towns, led in almost every village to construction of emergency living quarters and temporary 'crash' housing, even in places where there would probably have been no reason to locate when an initial decision was made, if the ordinary considerations used in an industrial society had been applied. The survival of the few remaining structures, mostly amid a heritage of rubble, is today leading back again to the same settlement structure as existed in districts in many of the smaller valleys, where only a third of the 1951 census population remained before the catastrophe. Curious incidents, such as the recovery of three bronze church bells unharmed from the ruins after the steeple crashed, have increased the determination of a pious, home-loving people to go on living in their traditional sites.[12] Pressures of time prevented us entering into discussion about this and inquiring into the possibilities of establishing up-to-date infrastructures.

If we take land-use changes as the starting point, such decisions have to be confronted very early, for private investors should not operate on their own and slip by all later planning. But land-use policy depends on verdicts from the experts about the relative risk in given areas – and experts decide only upon due reflection. Such experts generally come from outside the disaster area and are therefore less susceptible to the pressures of local public opinion. If they run into painful decisions about resettlement possibilities in particular areas, they lack the credibility accorded to local people. Consultations undertaken by outside relief agencies with regard to local decisions can be exemplified by the case of Cesariis di Sopra, a district of the commune of Lusevera.

The question is raised as to whether decisions that are sound from a natural science and emergency point of view are equally so from the social standpoint. The safety of an area in seismic or soil terms tells us little about the perhaps more important question of the supposed safety of a place judged by other environmental criteria. In the Tagliamento Valley, for example, the prefab housing areas were set up away from the threatening rockslide zone of cliffs, but the new settlements are planted right in the impact area of dense traffic arteries or placed, for example in Gemona, near torrential streams. A specific natural event first calls for expert judgments on this point, but related social science disciplines should in most cases also be consulted. How, for instance, should one proceed when land previously neither wanted nor used for agriculture is either immediately or gradually taken over for public use after a disaster? Many aspects of catastrophes have to be looked at within the framework of a given cultural situation. This applies especially to all problems regarding possible changes in building regulations.

Latin people are masters of stone construction, and turn all the more to native building materials because of the scarcity of wood in their lands. Ridgepole timbers carefully enshrouded in plastic on the half-vacated rubble sites show how prudently people deal with a precious building material even in the presence of disaster. Along the torrential streams of Friuli, a specially dangerous construction material is furnished by the rounded river boulders often used in place of quarry stone, which is a better but more expensive material. These rounded river stones, used with little mortar, were present particularly in the outer walls of older houses, mostly occupied by the older, poorer people. But even more modern houses proved to have skimpy roof-support frames, so that total collapse occurred in them as well. Furthermore, the double-rowed tile roofs sealed with mortar were very heavy, and so tended to fall through.

The prefabricated houses that were put up, notably those donated by other countries were, on the contrary, made almost entirely of wood or metal. Those prefabricated buildings that had already been erected before the second earthquake came through it very well because of their elasticity. They should have been able to give the local population confidence in a new construction technique.

Wooden houses, in the right spot, set among trees and cliffs, and at a suitable distance from their neighbors, can be very attractive. Northern lands where wooden buildings are traditional, such as Scandinavia, Russia or Canada, and also 'mediterranean' climatic zones such as California, make this evident. Stucco over a wood-frame house can convey the impression of a stone-walled house.

The process of educating the people of Friuli about earthquake-proof wood construction is taking place under difficult conditions, as can be imagined. Even where attractive wooden houses have been provided, they come out looking like barracks. This impression, now acknowledged in the use of the term 'baraccopolis', came into being because the buildings were set up very close together on standardized foundations to facilitate the arrangement of utilities. The time factor entered in here too.

The $-20°C$ cold spell around December 15, 1976 interfered with building the sewage system. Short service lines and the minimization of investment in the 'underground city', thus secured, however, meant reduced distance between houses and so produced this impression of monotony. The attempt to keep expropriated areas as small as possible also contributed to a high density in the crash housing zones (see Fig. 1.13).

Since nothing is more permanent than 'temporary' solutions, the authorities concerned had the following choice.

(a) To act in a 'deliberately callous' fashion and convert Friuli into a showplace for every conceivable sort of prefab housing, thereby providing rapid help by treating the problem as a temporary one and not laying permanent foundations. If they did so, they had to reckon with the fact that then they were giving the Friulians a warped picture of safe building methods using wood and the danger then arose that they would make their homeland so alien to them that it might no longer be regarded as such, and the local people would be thereby inclined to try to seek a solution by emigrating.

(b) They could try to save as much as possible of the old structures by steel and concrete preservation work (as the architect and landmark protection specialist R. Pirzio Biroli proposed and actually carried through in one village to provide an example). This would mean that identification with the destroyed living area would probably be achieved

Figure 1.13 The construction of prefab towns in the disaster area made only slow progress during the wintertime, because of slowness in the expropriation of land and development of building sites. Sometimes, the producers of prefabs or donors from abroad did not realize that parts of Friuli were mountainous and had masses of snow in winter. Chimneys and stoves had to be obtained and added.

sooner, but this would space the reconstruction out over a much longer period and exceed the timespan specified, to say nothing of the fact that preservation work could do nothing about houses that had been totally destroyed.

The alternatives presented in this case were either to proceed rapidly in the face of some alienating consequences, or to stick to the familiar way of building and proceed slowly. At the next higher level of decisions, this was linked with the integration of the individual buildings into a properly functioning and more attractive town. There are many works that portray the individuality of settlements in Friuli, especially those of Luciano Di Sopra, who even made use in some of them of Kevin Lynch's methods of analyzing perceptual qualities.[13]

Changes in the building regulations in Friuli should have, but had not, benefited from the experience accumulated in the USA since the Long Beach earthquake of 1933. And since so many Friulians had been employed abroad as construction workers, their experience should have been used in reconstruction. But their 'do it yourself' approach might threaten to seduce them into trying to save on building materials, finish and structural framing, which represent the largest costs in new construction. The builders could have taken an example from the adaptable prefabs made of wood, metal or synthetic substances. But the wishes of the overwhelming majority of Friulians run to a nice, solid, prestige-giving brick or concrete house. Accordingly, a part of our questioning was designed to get at changed ideas about building following the earthquake experience.

The fundamental problem of whether the city to be rebuilt will be different from the one destroyed in regard to greater attractiveness and functionality, is also evaluated variously among planners, architects, city administrators and the man in the street, because their time perspectives on the future and the expert insights needed for decision diverge too much. The conflict between the need to keep the presence of the prefabs from interfering with the reconstruction proper, which requires them to be situated at some distance from the destroyed centers (cf. the renewal plans for Osoppo and Venzone), and the need to make the town cores functional again represents a crucial problem. Since some sort of infrastructure has to be provided for the prefab quarter, the financial resources used up for this purpose in the baraccopolis reduce the funds available for regenerating businesses returning to the main part of town. Enterprises find no customers awaiting them there, but they have to invest there again forthwith to get reconstruction under way, for in fact it is they who make all the difference in revitalizing the core.

Since some Friulians began to restore their houses immediately after the May earthquake, thus using up their entire savings, only to have all their efforts at reconstruction again brought to naught by the second earthquake of September 15, 1976, a question of equity in their compensation is raised. Those who had stalled, in contrast to these early rebuilders, still had their savings, so that in this case an innovative reconstruction-mindedness was actually penalized. Our questionnaire research therefore also covered the problem raised by repeated destruction of property.

A change in accessibility was also brought about by the planning of the prefab areas. The classic settlement sites in Friuli are the alluvial cones (Gemona, for example) and moraines (San Daniele) bordering the glacial amphitheater, which means steeper slopes and less-usable tracts of land. Because of this, modern, mostly space-demanding manufacturing concerns have been located down in the floodplain along through-routes. Most settlements along the major traffic axis of Friuli, National Highway 13, running down through the Canale Valley from Tarvisio at the frontier to merge at Carnia with Highway 52 from the Tagliamento Valley, have grown out towards these roads that bypass towns, in order to profit from the traffic passing through, especially the tourists from Germany and Austria who head in summer for the Adriatic. Shopping centers are designed especially for the wants of tourists. A wide assortment of wines, liquors, fruit, shoes, clothing, handicraft wares and souvenirs is provided, even in tents and inflatable buildings. These buildings, mostly modern, withstood the earthquake well, and the businesses enjoyed a surge of demand for replacement items on the part of the local population. The commercial structure of the older town cores lying off the main routes, on the other hand, mostly serves the needs of the local folk. Many such businesses were completely wiped out. But the first emergency accommodations and 'tent cities' were set up on the unoccupied open land on the edge of the existing settlements and in places served by traffic, because community land (farmland earmarked for industrial development) was provided, and the property system was at first not radically modified in places that were in ruins, although Ordinance 43 afforded means to do so. This often caused the prefab quarters, and consequently the purchasing power of the local population, to move over to the through-highways, whose service stations, fast-food outlets, highway shopping centers, rest areas and industrial buildings with their corresponding billboards brought a decline in environmental quality when compared with the idyllic old town centers, which thereby were confronted with a further loss of vitality following their destruction. Compensation for such changes in locational relationships proved to be a difficult matter. We can especially learn from what happened in Friuli how to deepen our whole comprehension of individual and family circumstances associated with relocation. The issues include the evacuation (from the mountain districts) to a totally dissimilar milieu of hotel towns on the Adriatic; location of prefab tracts relative to original settlements; and the problems of assigning people places in the new prefab towns. Planners had the option of assigning people places in these facilities, either close to former village neighbors or in such a way as would deliberately create new communities out of former strangers who perhaps had become acquainted in the evacuation towns, or yet again simply in accordance with reconstruction plans already agreed upon. The choice between such options made Friuli a laboratory for empirical social research. Connections could be laid out between the urgent requirements of individuals and families and the economic circumstances within which these communities, the disaster area and the country were set. At the moment of greatest financial need, the local income sources failed while, extra-regionally, the catastrophe gave rise to growth impulses (as in the areas producing prefab housing). A redistribution of economic power took place, in which the innumerable victims of catastrophe were balanced off against people gaining something from the catastrophe, both within and outside the stricken area.

The advantages and shortcomings of a given economic system become particularly apparent in a time of crisis. Writers critical of the existing order therefore sometimes take almost perverse delight in likening 'earthquake' to 'classquake'.[14]

The foregoing details show that people responsible for planning at any level are confronted at the moment of catastrophe with a multitude of decisions that make great demands on their previous professional competence and their daily practical routine.

Disasters in certain cases can accelerate already present trends, but also may reverse them. Not only do they put the decision-maker in the situation of having, in the shortest possible time, to make decisions of the widest scope for the future, which, were it not for the disaster, might have been spread out over a much longer period, and which might have been qualitatively dissimilar; they also find themselves required to consider anew, confirm or reject most of the decisions made sometimes generations before by others, the effects of which still hold in the area's spatial framework. This starts with the basic decision of whether a destroyed settlement should be re-established on the same site at all, and if so when, and what functions it should assume among other settlements intact or also destroyed. The consolidation of districts has to be considered, as in Alaska, because of changed ideas about the most efficient population size of districts for a given function in relation to optimal distribution of support facilities. On the other hand, the degree of sensed unity of the population concerned, such as is so strongly marked in Friuli's encapsulated valleys, has to be taken into consideration. The entry of the evacuated children into 'valley community' school classes in the temporary facilities of Adriatic cities was fully in accord with the feeling of belonging together, a very important matter indeed given the necessity of living so close together in the vacation apartments and later in the prefab quarters (see Fig. 1.14).

More than in other catastrophes that have been subject to research, the case of Friuli should be seen in its regional and inter-regional dimensions. Many authors tend to envisage a disaster as occurring only at one point. Characteristically, they speak about 'the city' (e.g. of San Francisco, Anchorage, Managua etc.), because recent or historical experiences of catastrophe conformed to a point scale. Friuli, however, was affected over an area of 4800 km². The decision-makers responsible for reconstruction had to solve problems of social policy with all their implications for equity. There are notable differences between European and American 'planning' institutions, especially in the sector of public versus private initiatives. And there is, of course, no formula that could ever answer questions of the future prospects of

Figure 1.14 In the school for evacuated children in Grado, a drawing teacher asked the children to paint the traumatic experience of the catastrophe. One child painted a giant without a head, stamping his way through the country. Wherever he steps, the earth quakes!

an area so large. What ought to become of Friuli? Should it be:

(a) a quickly-covered stretch of highway to the Adriatic beaches?
(b) a cautionary educational exhibit?
(c) a trade fair for prefabricated housing?
(d) a once more viable region of Italy with an industrial speciality it has learned from its own experience: earthquake-proof buildings, a laboratory for new construction techniques?

These and similar questions occupied the Munich research team as it began its first investigations in the disaster area in the winter of 1976–7. The emergency administrators, evacuation specialists, engineers and geologists, military personnel and rescue units had given their best. Much had to be decided quickly – too quickly, indeed, it seemed to us. Faced with such a large amount of expert help, had the people involved even had a chance to articulate their problems from their own viewpoints?

It would have been unpardonable, in the light of the material help of so many ordinary people from the countries neighboring Friuli, and the spontaneous gratitude of the earthquake victims for this help, if applied social geography were to decline to take up problems whose solution could be seen as an obligation of science and a contribution toward solidarity within the European Community.

1.4   Problems created for the people affected

Earthquake measurements are normally expressed in physical units such as points on the Richter scale. The *Earthquake Information Bulletin* of the USGS (1976) gives an intensity of 6.5 R for the Friulian quake and Italian sources evaluate the May earthquake at 6.4 R and the one in September at 6.1 R.[15] The modified Mercalli intensity scale is based on actual observation and makes regional differentiations according to the principle of concentric rings. More to the point, however, is the measurement of social disruption. Structural measurement data may be derived from the number of buildings destroyed or unsafe in varying degrees. They tell us nothing about whether all the buildings were occupied, and what help was needed for how many persons. The effect on people, however, could really only meaningfully be gauged in measurement units provided by the social sciences. These are only very poorly developed.

At the same time, questions can arise in such cases about the reliability of the measurements used. Indices of numbers of people left homeless or evacuated are collected under a crisis situation and reflect it in their quality. Sometimes they are quite consciously increased, and they can be colored by policy considerations in the early stages of a catastrophe; for instance, a high number of homeless reported will supposedly emphasize the extreme need of a district and attract scarce resources to it. During the evacuation operations in Friuli, such varied numbers were given for people declining to leave the area that they could only be explained on the basis of the individual circumstances in each district. After the process of bringing the people back had been completed, refusals on various pretexts to return cast doubt on some of the reasons for the whole operation. Data for small districts are probably more reliable than those for the larger cities. Duplications are apparent in the enumeration, for instance, when the figure given for the homeless from the May earthquake plus that for the homeless from the September one add up to more than the total population.

In several early attempts, such as the decrees of May 18[16] and May 20[17] 1976 defining the disaster zone, three categories of effects were set up. The latter decree differentiates between destroyed communes, badly damaged communes and damaged communes (Table 1.4, Fig. 1.15). The three categories thus embrace 41, 45 and 33 communes respectively. The special number of *Il Messaggero Veneto* – 'Documento', December 31 1976,[18] p. 4 – gives, however, 41 districts of category 1, 43 districts of category 2, but 51 of category 3. A study group of the Autonomous Region of Friuli–Venezia Giulia directed by Luciano Di Sopra, a professor of architecture, calculated the relative damage in the three zones as follows.[19]

Zone 1, with 55 000 inhabitants and the following characteristics: only 10.2 per cent of the population of the affected region as a whole, but 81 per cent of the fatalities, 68 per cent of the injured, and 62 per cent of the homeless.

Zone 2, with 75 000 inhabitants and the following characteristics: 14 per cent of the population of the affected region as a whole, 18.5 per cent of the fatalities, 21.1 per cent of the injured, and 35.5 per cent of the homeless.

Zone 3, with 259 000 inhabitants and the following characteristics: 46.6 per cent of the population of the affected region as a whole, 0.3 per cent of the fatalities, 7.6 per cent of the injured, and 2.5 per cent of the homeless.

Finally, the remaining zone of the Provinces of Udine and Pordenone, with 156 000 unaffected inhabitants (29.2 per cent).

It will be shown subsequently that some communes were put into the wrong categories.

Later corrections were made in some instances, but could not always be consulted. Unfortunately, different standards of measurement were applied by the central government, the region of Friuli–Venezia Giulia, and the two affected provinces of Udine and Pordenone on occasion. The questionnaire administered by us, which is referred to in Section 2.2, used the demarcations in force before April 22.

It should also be pointed out that a lesser amount of damage also occurred in neighboring Slovenia and in the Gailtal of Carinthia. The fact that our discussion stops at the Italian border is due to data limitations. The seismo-tectonic map (Fig. 1.10), however, shows particularly the potential threat to the areas of Villach and Belluno.

Our research team, once it began its fieldwork in the disaster area, soon realized that the delimitation proposed was not representative of the real extent of social disruption, that some of the hardest hit communes had been ignored by the Italian authorities and that, on the other hand, some only slightly affected communes had been placed in a much too high damage category.

Therefore we tried to use, instead of the usual Richter and Mercalli scales,

Table 1.4 The three categories of commune.

(a) Comuni disastrati (destroyed communes) of the provinces of Udine (29) and of Pordenone (12), namely:

Province of Udine		Province of Pordenone
Amaro	Moggio Udinese	Castelnuovo del Friuli
Artegna	Montenars	Cavasso nuovo
Attimis	Nimis	Clauzetto
Bordano	Osoppo	Fanna
Buia	Pontebba	Frisanco
Cassacco	Ragogna	Meduno
Cavazzo Carnico	Resia	Pinzano al Tagliamento
Chiusaforte	Resiutta	Sequals
Colloredo di Montalbano	San Daniele di Fr.	Tramonti di Sopra
Faedis	Taipana	Tramonti di Sotto
Forgaria	Tarcento	Travesio
Gemona	Trasaghis	Vito d'Asio
Lusevera	Treppo Grande	
Magnano in Riviera	Venzone	
Majano		

(b) Comuni gravemente danneggiati (badly damaged communes) of the provinces of Udine (39) and of Pordenone (6), namely:

Province of Udine		Province of Pordenone
Ampezzo	Povoletto	Andreis
Arta Terme	Premariacco	Arba
Buttrio	Preone	Maniago
Cercivento	Pulfero	Montereale Valcellina
Cividale del Friuli	Ravascletto	Spilimbergo
Comeglians	Raveo	Vivaro
Dogna	Reana del Roiale	
Enemonzo	Remanzacco	
Fagagna	Rive d'Arcano	
Lauco	S. Pietro al Natisone	
Ligosullo	Socchieve	
Malborghetto	Sutrio	
Martignacco	Tricesimo	
Mereto di Tomba	Tolmezzo	
Moimacco	Torreano	
Moruzzo	Treppo Carnico	
Ovaro	Verzegnis	
Pagnacco	Villa Santina	
Paluzza	Zuglio	
Paularo		

(c) Comuni danneggiati (damaged communes) of the provinces of Udine (29) and of Pordenone (4), namely:

Province of Udine			
Basiliano	Forni di Sotto	Prepotto	Tavagnacco
Campoformido	Grimacco	Rigolato	Udine
Corno di Rosazzo	Lestizza	S. Giovanni al N.	
Coséano	Manzano	S. Leonardo	Province of Pordenone
Dignano	Pasian di Prato	S. Vito di Fag.	Cordenons
Drenchia	Pavia di Udine	Sauris	Fontanafredda
Flaibano	Pozzuoli del Fr.	Savogna	Pordenone
Forni Avoltri	Pradamano	Stregna	Sacile
Forni di Sopra	Prato Carnico	Tarvisio	

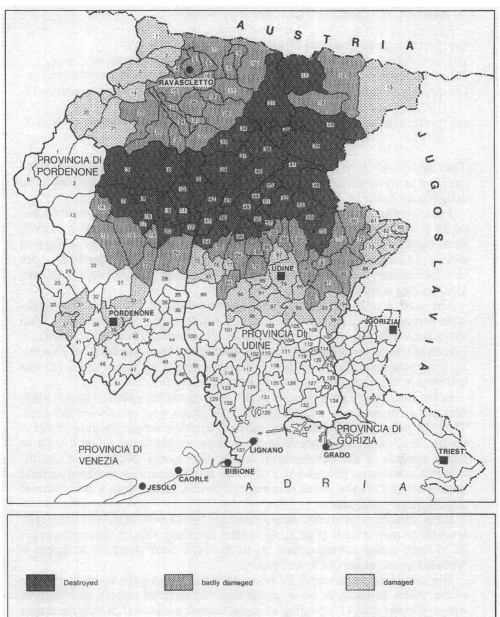

Figure 1.15 Classification of the disaster area into three classes: 'destroyed', 'badly damaged' and 'damaged'.

a whole series of social indices in order to define the area of maximum damage level, to which the greatest amount of relief should also accordingly be directed. We chose for this purpose five variables:

(a) percentage of dead out of total population of commune;
(b) percentage of homeless on May 6, 1976 within a commune;
(c) percentage of homeless on September 15, 1976 within a commune;
(d) percentage of commune population that was evacuated by December 18, 1976;
(e) percentage of people in the commune living in prefab housing on May 1, 1977.

They were added up, with no weighting, and the highest-ranking 30 were assigned corresponding index numbers; places ranking beyond 30 were all assigned an index of 30.

The commune of Lusevera can be used as an example; with an index value of 32 it took first place in respect of effect intensity. This figure is derived from the fact that Lusevera ranked ninth with respect to dead, seventh in number of homeless in May 1976, first in homeless in September 1976, 13th in total evacuated in December 1976, and second in prefab dwellers in May 1977, giving a total of 32.

The same was done with all the communes listed in Table 1.5. The point-scale was carried out to over 250 and, as a result, took in 60 communes. With the 52nd commune, the least affected of the category of 'comuni disastrati' (which of course included only 41 communes) had been reached (i.e. Faedis). Eleven communes must therefore have been harder hit than Faedis, but not included in the category 'comuni disastrati'.

In the interests of providing an easily understood formula for Italian planners and administrators, strict statistical procedures were not always followed. Thus the points in Table 1.5 were determined by a simple addition of ranks on the five variables, rather than by a more sophisticated method such as factor analysis or the development of standard scores. So, too, the classification of communes followed an essentially intuitive procedure based on field inspection and breaks in the sequence of points, rather than a more formal classificatory procedure.

If the commune of Resia, under-rated for this reason, is moved from type level III to type level II (Fig. 1.16), it then faithfully reflects the earthquakes as to their social consequences, in so far as a clear decrease in effect is revealed from center out to periphery.

The communes of category IV occur exclusively along the southern margin of the major catastrophe zone, exhibiting two definite eastern and western wings of lesser effect (according to social science measures), which are stages apparently missing in the transition from the epicenter to the north. However, such a transition belt is present there too. It becomes evident when comparisons are made with data from the southern border communes of the 'zona disastrata', for then it shows up in the fact that the perception of degree of impact is displaced asymmetrically towards the headquarters of the people responsible for officially determining intensity of effect, namely the prefectures of Udine and Pordenone. More simply put, misery close by is taken more seriously than that more distant.

Table 1.5 Intensity of impact. Ranks of communes according to five criteria and following two methods. Communes above the dashed line without * and with ! should be upgraded. Those below that line with * should be downgraded.

Rank	Points	Commune	CD	CV	Key
1	32	Lusevera	*	19	
2	49	Trasaghis	*	17	
3	67	Tarcento	*	16	
4	71	Venzone	*	16	
5	73	Gemona	*	16	
6	73	Cavazzo Carnico	*	15	I, 10 communes
7	76	Bordano	*	15	
8	77	Montenars	*	14	
9	78	Osoppo	*	15	
10	78	Forgaria	*	15	
11	85	Pinzano	*	13	
12	89	Nimis	*	14	
13	92	Vito d'Asio	*	13	
14	95	Buia	*	14	
15	97	Dogna	!	11	
16	101	Chiusaforte	*	13	
17	101	Magnano di Riv.	*	14	II, 13 communes
18	102	Moggio Udinese	*	14	
19	103	Amaro	*	13	
20	111	Cavasso nuovo	*	11	
21	114	Castelnuovo	*	11	
22	115	Resiutta	*	11	
23	122	Artegna	*	14	
24	127	Sequals	*	7	
25	128	Resia	*	13	
26	133	Tramonti di Sotto	*	9	
27	137	Attimis	*	10	
28	137	Pontebba	*	8	
28	139	Treppo Grande	*	9	
30	142	Majano	*	9	III, 13 communes
31	143	Colloredo		10	
32	151	Villa Santina	!	6	
33	156	Clauzetto	*	7	
34	161	Tolmezzo	!	6	
35	164	Preone	!	6	
36	174	Ragogna	*	9	
37	184	Travesio	*	4	
38	185	Meduno	*	4	
39	192	Reana del Roiale	!	6	
40	193	Raveo	!	4	
41	197	Arta Terme	!	2	
42	197	Tramonti di Sopra	*	7	
43	199	Fanna	*	6	
44	209	Povoletto	?	2	
45	216	Verzegnis	?	3	IV, 16 communes
46	217	Taipana	*	8	
47	222	Enemonzo	?	1	
48	225	Zuglio	?	4	
49	226	Frisanco	*	5	
50	229	Cassacco	*	3	
51	231	San Daniele	*	4	
52	235	Faedis	*	4	

Table 1.5 *continued*

Rank	Points	Commune	CD	CV	Key
53	239	Tricesimo			
54	239	Spilimbergo			
55	240	Pagnacco			
56	241	Socchieve			
57	246	Torreano			
58	247	Ligosullo			
59	252	Remanzacco			
60	253	Ravascletto			

CV, combined value; cd, comuni disastrati; !, definite misplacing; ?, possible misplacing.

In order to compensate for the distortion of the 'perception surface', it was therefore proposed to upgrade nine communes in the north and two in the south, and to downgrade six communes in the south. Figure 1.17 should be compared to Fig. 1.16; correction would be particularly needed if the official division of Friuli's communes into three categories of effect were to be actually involved in decision-making procedures. The 'Decreto del Presidente della Giunta No. 01009' of April 22, 1977 brought about such a correction in favor of the communes of Spilimbergo, Tolmezzo, Tricesimo and Villa Santina. In the cases of the communes of Villa Santina (32nd place on Table 1.5) and Tolmezzo (34th place), the correction accords with the new boundary proposed by the author (Table 1.5). On the other hand, Tricesimo and Spilimbergo only show up in 53rd and 54th places. In contrast, Dogna, in spite of holding 15th place on the effect scale, gets forgotten again.

In the lower effect categories, too, shifts took place. The 'Decreto del Presidente della Giunta No. 0351' of February 8, 1977 inserted the commune of Porcia, the western neighbor of the provincial capital of Pordenone (commune index number 38) into the group of the 'comuni danneggiati', with an eye exclusively to the factory site of the well known Zanussi electrical household appliance concern.

Such on-again, off-again decisions show that we still know much too little about the typical course of events in catastrophes. They appear, particularly in a really fragmented, small-scale Europe, to be unique events to be dealt with each time on an individual basis. Within divided Europe, nobody considers what happens in Skopje, in Sicily, in Rumania, on the Macedonian Adriatic coast of Yugoslavia, and in Friuli, within a single systematic framework. It seems harder to do so there than it might be in the United States, for example. The regional structures are too diverse, as between the Anatolian hinterland of Turkey and some Central European area such as Friuli, so much more highly developed in comparison.

For this reason, the next chapter will be an attempt to survey the present state of hazard research, using earthquake hazard as an example, and to relate what has been learned in Friuli concerning that field. It is hoped that this will permit the application of more general concepts to this particular case and that, on the other hand, theory will be enriched by what was learned in this special instance.

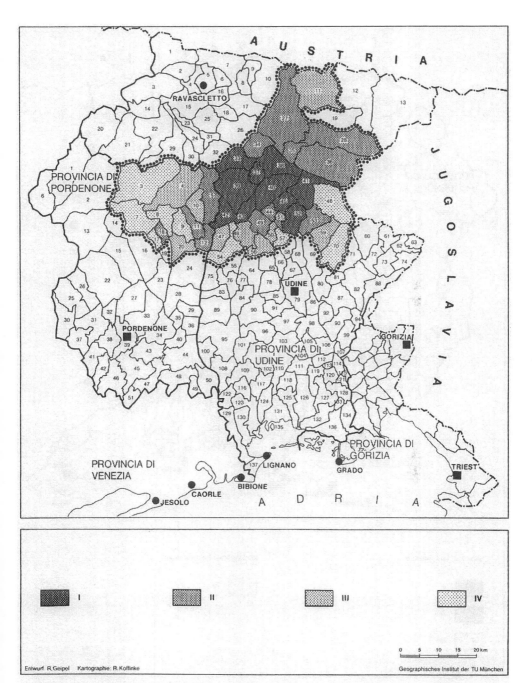

Figure 1.16 Subdivision of 'destroyed communes' into four categories.

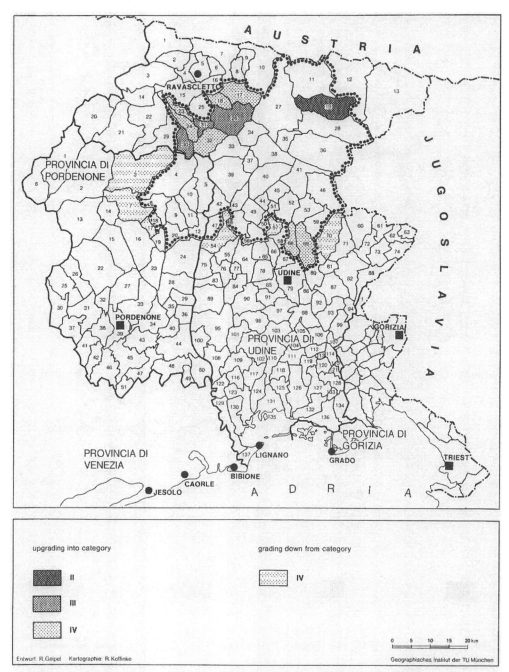

Figure 1.17 Proposal for a new definition of the category 'destroyed communes'.

Notes

1 Turner, R. H., J. M. Nigg, D. H. Paz and B. S. Young 1979. *Earthquake threat. The human response in southern California*. Los Angeles. This refers on page 9 to Friuli, together with Guatemala and the People's Republic of China among the events of 1976.
2 Melchoir, G. and L. Di Sopra (eds) 1976. *Prime ipotesi di intervento nelle aree colpite dal sisma del 6 Maggio 1976*. Majano.
3 Valussi, G. 1971. Il fenomeno migratorio in Friuli far i processi di deruralizzazione e industrializzazione. In *La realta sociale di una regione in fase di sviluppo*, 104–26. Udine.
4 Valussi, G. 1971. L'emigrazione nel Friuli–Venezia-Giulia. In *Encyclopedia monografica del Friuli–Venezia-Giulia*, vol. II, 853–928. Udine. Pagani, B. M. 1968. *L'emigrazione friuliana dalla meta del secolo XIX al 1940*. Udine.
5 Valussi, G. op. cit., p. 903.
6 Wagner, P. Personal communication about the Friulians in Vancouver after the 1976 disaster.
7 Boulding, K. 1976. *The role of catastrophe in evolutionary dynamics*. Symp. Am. Assoc. Adv. Sci., Boston, February 21.
8 Severi, F. S. 1977. Il richiamo del commissario straordinario nel Friuli terremotato. *Le Regioni. Riv. Ist. Studi Giuridici Regionali* IV, 44–73.
9 Di Sopra, L. Personal communication, extended by the author and confirmed by information from Karl Folladori of the Bavarian Red Cross field party.
10 Regione Autonoma Friuli–Venezia-Giulia, Consiglio Regionale 1976. *Disposizione statali e regionali relative ai provvedimenti di primo intervento e di ricostruzione nelle zone del Friuli colpite dagli eventi sismici dell'anno 1976*. Udine.
11 See note 3 above.
12 Davis, N. Y. 1970. The role of the Russian Orthodox Church in five Pacific Eskimo villages as revealed by the earthquake. In *The great Alaska earthquake of 1964*, vol. 6. *Human ecology*. Washington, DC.
13 Di Sopra, L. 1970. *Le Prealpi Giulie. Ricerca per un piano comprensoriale* (3 vols). Udine. Di Sopra, L. undated. *Spazio urbano e prezzo del suolo. Zona collinare Friulana*. Udine.
14 O'Keffe, P., K. Westgate and B. Wisner 1976. Taking the naturalness out of natural disasters. *Nature* **260**.
Jäger, W. 1977. Katastrophe als sozialer Prozess. Alternativen zur gegenwärtigen Katastrophenforschung. *AIAS-Informations* 1/2.
15 Ambrasey, N. N. 1976. *The Gemona di Friuli earthquake of 6 May 1976*. Restricted Technical Report FMR/CC/SC/ED/76/169, p. 78. Paris: Unesco.
Bartole, R., G. Bernadis, F. Giorgetti, D. Nieto and M. Russi 1976. Earthquake catalogue of Friuli–Venezia-Giulia region. *Boll. Geofis.: Teor. Applic.* **XIX**, 308.
United States Geological Survey 1976. *Earthquake Information Bulletin* 8(5), 30.
16 See note 10.
17 Decreto no. 0714, p. 61. Other decrees of importance are: Decreto no. 02286, December 9, 1976; Decreto no. 0351, February 8, 1977; Decreto no. 01009, April 22, 1977.
18 *Messaggero Veneto* 1976. Special edition, 'Documento', December 31, p. 2.
19 Di Sopra, L. 1976. *Relazione di Sintesi. Sisma del 6 Maggio*, 3–5. Udine: Regione Autonoma Friuli–Venezia-Giulia, Segreteria Generale Straordinaria.

2 Learning from Fruili

2.1 The relevance of hazard theory for the case of Friuli

Many kinds of natural hazards have become the subject of social geographic research since the 1950s, especially in the United States. Worldwide comparisons have been made recently.[1] Among them, earthquakes particularly merit the attention of a social science oriented geography. This is especially so because, until recently, no way of prediction and early warning was available. Unlike the case of weather phenomena such as tidal waves, floods, hurricanes, tornadoes or typhoons, or of forest fires, no advance warning can be obtained on the basis of calculated trajectories, time of impact, etc. in order to facilitate evacuation measures. The geographer–planner, instead, must deal with long-range provisions for emergency relief in zones of high-settlement density. This is because, unlike drought, cold waves and storm disasters, which generally mainly affect the crop harvests of farmers living dispersed over an area, earthquake damage is most severe and most extensive precisely in settlements where the activities of society are most complex and subject to disruption.

Furthermore, earthquakes can trigger secondary hazards such as the flooding caused by the bursting of dams, landslides, avalanches, fires, and epidemics, and consequently increase the dangers inherent in hazards created by man, e.g. atomic power plants, refineries, oil pipelines, or structurally hazardous traffic facilities such as harbors, bridges, tunnels etc. Increasing urbanization – involving ever greater building densities in the inner-city high rise zone, with its always bolder, speculative, prestige-oriented construction – signals a gradual trend toward steadily increasing susceptibility to trouble for an ever greater portion of mankind.

Even in Central Europe, which has previously been largely spared from natural catastrophes, we must also pose the behavioral question of how human beings react to unpredictable natural events. The Dutch and North German flood catastrophes have shown that we have no grounds for supposing ourselves to be secure within a world protected by technology. Technology itself, in the form of 'manmade hazards', can be more of a threat to life than natural hazards.

What happened in Friuli can be compared, for the areal extent of damage, loss of life and property destruction, with the Dutch flood disaster of 1953[2] and the North German floods of 1962. In Friuli, an area of 4800 km^2 was affected,[3] in the Netherlands 1600 km^2. The maximum number of persons evacuated or homeless was around 100 000 in both places. At the then current purchasing power, 860 000 000 Dutch florins worth of damage was the equivalent of the estimated \$2 billion of direct damage to buildings in Friuli. Taking into account the cessation of production, the downgrading of economic life in general to a lower level, and the damage to cultural treasures that is so difficult to estimate, the net balance of destruction overall may come to some 4400 billion lire[4] or \$6 billion for the region.

Spatially significant activity by government on a major scale usually follows any large disaster. It is therefore mostly planning bodies that document catas-

plish in areas of intense speculative pressures and suspicion-laden restrictions on the autonomy of public planning authorities, as in Japan or California than it is in systems with centrally directed economies and almost unlimited state authority. Under 'free-market' mechanisms, e.g. insurance premiums graduated according to the risk connected with particular building characteristics (type of structure, materials used, age of structure, location with respect to especially unsafe lines of cleavage, angle of slope, character of the subsoil zone), hardly anything can be done, in the absence of compulsory insurance, about the rehabilitation of existing building stock.

While about a million earth shocks occur each year, with an average annual frequency of about 20 strong quakes and a couple of catastrophic events (based on the period 1900–64), data on recurrences can be reasonably well authenticated only in areas with good documentation.[12] Friuli, an area of old civilization, is far ahead of California in this respect. The ability to make *regional* predictions of earthquake hazard is already very well developed in such areas. Maps now exist, not only at world scale but for certain countries, showing the likelihood of earthquake hazard.[13] Actual timing and intensity of a quake, on the other hand, are far harder to predict. Japanese official agencies began an early-warning service in 1966. Chinese successes and failures in earthquake prediction were publicized during the Unesco Conferences of February 1976 and April 1979 in Paris,[14] and gave rise to more intensive efforts in this domain of research in other countries.

The willingness to take account of earthquake hazards, of course, sometimes tends to develop by fits and starts. After the San Francisco earthquake of April 18, 1906 (which in combination with the great fire that followed it destroyed 28 130 structures) the term 'Great Fire' soon came into vogue and banished the earthquake that had been its cause to the borderlands of public consciousness,[15] while at the same time a feeling of security arose because the release of earth tensions should supposedly be followed by an interval of 50 or 100 years of calm. Daniel Burnham's 1905 development plan for the city, offering the prospect of a more open layout with wide streets and broader squares, was frustrated after the earthquake by the prevailing speculation in real estate, even though the best possible circumstances were then at hand for putting it into effect. Even so, up until the First World War, no newly constructed buildings exceeded seven storeys in height. Instead, the Central Business District (CBD) expanded horizontally by 44 per cent between 1906 and 1915. The first buildings of more than 25 storeys were built in the mid-1920s, and it was only in the mid-1970s that they went beyond 60 storeys.

There had never been any need for high-rise construction in Friuli. Even in Udine, the provincial capital, most buildings do not exceed 10 storeys. Since the earthquake, their upper floors have become harder to rent. Even so, a 10-storey Central Hospital was built in Gemona, and now for safety reasons this new building, never having been occupied, had to be demolished in June 1979, after sustaining extensive damage (Fig. 2.2). This fact shows clearly how secure the people and officials in Friuli had felt, despite a history of earthquakes documented over centuries.[16]

Twenty-seven years elapsed between 1906 and the next time the American public's consciousness was jolted: the Long Beach earthquake of 1933 led to changes in the building codes. Thirty-one years later, the great Alaskan earthquake of 1964 released a flood of activity. For instance, the whole 1400-km

Figure 2.2 When the brand new, but unsafe, Central Hospital in Gemona was dynamited in June 1979, the myth that Friuli is not a seismic risk area was demolished. That the hospital had been built in Gemona at all (and in the form of a high-rise building!) shows that not only the population but the officials too had totally suppressed the idea of an earthquake.

length of the San Andreas Fault was then recorded in detail on topographic base maps for the first time, and a dense network of seismographic observation stations was set up.[17] The effects and consequences of the Alaskan earthquake were studied and described by a mixed commission set up on the orders of President Johnson. An interdisciplinary team made up of geologists, seismologists, geodesists, hydrologists, biologists, oceanographers, coastal protection specialists, engineers and social scientists produced an eight-volume document, of which the volume on human ecology deserves the special interest of geographers.[18]

The next decisive stage in the development of public consciousness was the San Fernando earthquake of February 9, 1971. It did not take place, as the Alaskan earthquake had done, on the periphery of the United States, but rather on the edge of one of its densest population concentrations. It furthermore represented a test for the stricter building specifications adopted after the 1933 Long Beach earthquake. The San Fernando earthquake was investigated and documented more summarily than, but in a similar fashion to, the Alaskan earthquake by an interdisciplinary commission.[19]

The foregoing remarks emphasize, with examples taken from the United States, that interest focuses on the earthquake theme periodically, at intervals of 25–30 years, because obviously in a country of subcontinental scale earthquake catastrophes will recur somewhere within such a timespan and evoke official reactions. Since San Fernando, this awareness of the problem has been accentuated and kept alive by disasters outside the United States given detailed coverage by the media, for which the country has usually contributed extensive relief aid. The number of interdisciplinary symposia and conferences where social scientists, Earth scientists and engineers make contributions is increasing. Legislative measures in particular states, notably California, have become stricter.[20] In September 1973, the earthquake research programs of the National Oceanic and Atmospheric Administration (NOAA) and the US Geological Survey were combined.[21] Land-use regulations, emergency plans and evacuation plans are required as 'Seismic Safety Elements' by the State Planning Law revision in 1971. With Senate Bill 1729 (Alquist), a Seismic Safety Commission (SSC) was established on July 1, 1975.[22] The present writer's collaboration with some American colleagues traces back to one SSC proposal; specifically, the one calling for 'an immediate investigation of earthquakes that have taken place outside California'.

The possibility of prediction became a new element in a worldwide discussion of earthquake hazards from 1974 onwards. Visits by American seismologists to the People's Republic of China in October 1974 disclosed a surprisingly high level of prediction techniques. The conferences in Paris in 1976 and 1979 linked this growing knowledge with the possible socioeconomic consequences of disaster warnings.[23] There has hardly been any such development of warning systems in European countries.

The social science contributions to hazard research are also just in their early stages. It was the events in Friuli that brought Central Europeans across the threshold of awareness, and we shall have to test how long it takes that population to show in its decision-making that it has learned something from the catastrophe – and how rapidly the consciousness of risk is lost again. (In our questionnaire, an attempt was made to put this learning process into operational terms by way of a question about preferences for various construction materials.) Since earthquake disasters in Europe have hardly been systematically studied yet, it was necessary to draw on examples from the United States for confirmation of the instability of readiness to recognize the risk of earthquakes. Figure 2.3, drawn from White and Haas (1975),[24] is an attempt to show the connections between the natural event, the secondary effects it sets off, and their impact on society in the stricken area. Any new catastrophe that occurs will introduce new components into this scheme, or bring out particular connections among its elements, as indeed was the case with Friuli. The very strong aftershock of September 15, 1976 and the series of individual rockfalls that took place over a one-year span, for example, lend emphasis to the line connecting 'aftershocks' to 'anxiety'. Rockfalls and the lack of water resources were especially prominent among the secondary effects produced. Structural damage in this area of artistic treasures was magnified in the cultural dimension. Social disruption was accentuated by the evacuation to Adriatic coastal cities. The time at which the catastrophe oc-

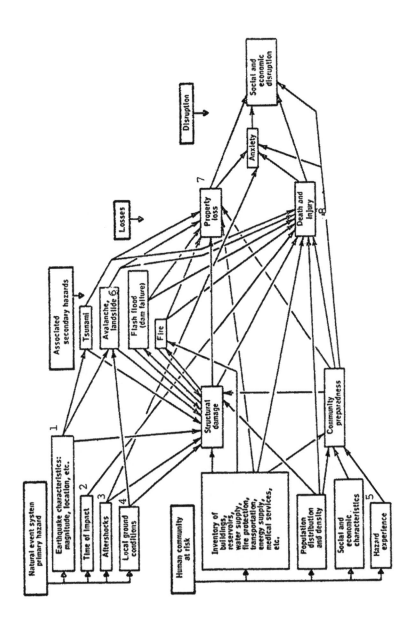

Figure 2.3 Some relationships of the impact of earthquakes on human social systems (White & Hass[24]). The numbers refer to some details of the Friuli earthquake: 1 = 6.5 R (May 6, 1976), 6.1 R (September 15, 1976); 2 = 9 p.m., warm temperature; 3 = about 400 aftershocks between the two major earthquakes; 4 = higher damage on alluvial fans; 5 = practically none; 6 = rock-falls blocking escape routes; 7 = estimated at $4.75 billion; 8 = 1000 dead, 2400 injured.

curred – 9 p.m. – and the construction materials employed in the old quarters of the towns were responsible for high casualty rates. A more favorable influence was the very hot weather which kept numerous people outdoors in their gardens.

The application of the results of American studies to the Friulian earthquakes of May 6 and September 15, 1976 can help make up for the paucity of European hazard research,[25] but, on the other hand, what we have found out here can extend our knowledge of the social–spatial consequences of catastrophe. The mobility processes set off by catastrophe, above all, and the planning problems involved in reconstruction present new research problems for applied social geography. Among these is the heightened consciousness of planning needs in areas of comparable risk and similar social circumstances, but sometimes with different official policies with regard to emergency measures. The Friulian case may set a notable example of bureaucratic structure for planning, involving temporary installation of a crisis manager in the person of the *Commissario Straordinario*, that ought to be carefully considered in countries of relatively weaker central authority, such as the United States. We may, on the other hand, suppose that citizen participation and the spontaneous organization of aid and reconstruction measures in the USA go back to a more mature tradition of self-rule. The author conceives his role as that of a two-way interpreter, trying out aspects of American research on a tragic case in neighboring Italy for use in German research, and bringing together methods from 'over there' with problems and data from 'here'.

2.2 The need for research and the design of our study

The material about the situation of the stricken population of Friuli that has been introduced so far was derived exclusively from official data of the regional authorities and Emergency Commissioner. Could this material be trusted?

We can see in the case of delimiting 'comuni disastrati' how local authorities may wear blinkers. Can we be content with proclamations and political declarations beamed toward potential voters? Should we not look for answers to our questions among the affected people themselves?

We decided, in fact, to interview the people living in the barrack settlements in the heart of the disaster area.

The questionnaire, reproduced in translation in the Appendix on p. 195, was aimed at the following aspects of the problem.

How will the population react to different regional economic alternatives in reconstruction? Such alternatives, if we follow growth-pole theory, could have involved in the most extreme case not bringing large portions of the population back to their home communes from the coastal towns after the evacuation, but concentrating the people where possible in 'new towns'. Outside the destroyed area and away from the zone in which there is a risk of repetition of seismic disasters, away from the mountain districts that have no adequate infrastructure, one or several such 'new towns', perhaps along the development axis of Pordenone–Udine–Gorizia–Trieste or simply just 'New Udine', would be conceivable. A building program able by May 1, 1977 to

43

afford emergency accommodation in temporary housing to 65 000 persons would have been able to accomplish this, not dispersed over 95 communes but as implementation of such planning of growth poles at one or a few places. Even if the regional plans for the provinces of Udine (309 711 m² for 23 955 persons) and Pordenone (33 587 m² for 2782 persons) had conflicted with such goals, the Commissario Straordinario's planners had at their disposal a quantity of (for Udine) 357 551 m² for 25 955 and (in Pordenone) 63 937 m² for 5386 persons. Would such 'new towns' have been accepted? Would they persist in the future, especially when the container-dwellings turned into sauna baths the next summer, and in spite of the fact that the prefabs were so clearly tagged as 'temporary' in the very beginning?

Question 17 of the questionnaire attempts to get an answer to this from the people living in prefabs. It focuses on 'readiness for urbanization' and deals with two different-sized cities – Udine and the regional capital of Trieste. In some cases, when permanent city residence seemed conceivable to respondents (mostly young ones) Trieste, which was disliked, was crossed out.

What will the population try to do in the near future to improve its present situation? Question 15 attempted to measure attitudes toward reconstruction (optimism/pessimism). It tested for the alternatives:

(a) get to work right away on rebuilding the houses (question 15, part 1); or
(b) stay in the emergency relief accommodations (question 15, part 5);
(c) the likelihood that portions of the population (cf. questions 2,3,4, and 12 on family status, age and source of income) will consider the temporary arrangements as a permanent solution (indicated in the case of Val Belice).

The readiness, alluded to already in question 17, to settle in a more urban commune or one other than the one of origin was tested by question 15 in three variants:

(a) disposition to mobility within the region (part 2);
(b) disposition to permament inter-regional mobility (part 3);
(c) disposition to intraregional mobility with intention to return (part 4).

A response along the lines of part 4 of question 15 would indicate a strengthening of the already manifest, typical migration behavior, following which many Friulians seek work abroad but at retirement age return to their home communes – a matter also inquired into by *question 9*, concerning property holding in the home commune (many such old-age provisions were lost in the earthquakes, and crisis housing was called into use for this end when available).

Questions 5, 6, 7, 8, 10 and *11* sought to go deeper into this central problem and, at the same time, to determine whether on account of:

(a) loss of work place caused by the earthquake (question 6);
(b) unemployment in general (question 10, part 4);
(c) long absence from the home commune because of work elsewhere in Italy (question 10, part 2);

(d) long absence from the home commune because of work outside the country (question 10, part 3);

(e) change of work place (question 11);

the attachment to the place of residence is loosened, especially in the case of:

(a) loss of a near relative (question 5), or of

(b) the family home (question 7), or of

(c) a non-agricultural business (question 8).

Did the peoples' ideas about construction change? A third part of the questionnaire applies to the learning process with regard to choice of building materials that both earthquakes might have initiated (question 16). The classic kind of construction with blocks of stone, rounded boulders from the streambeds, but at the same time with all too sparse quantities of reinforcing steel and thin cement can be held responsible for the high number of casualties. Will the people draw conclusions from this that would be important for the building industry? Are there the beginnings of acceptance of the wooden houses present in other Alpine regions, or have the unpopular wooden barracks extinguished this possibility? Many Friulians are all too used to living in barracks as a result of being abroad so long as construction workers. They could get used to living in these barracks forever.

What was the effect of the repetition of the earthquake? Finally, another issue is touched on in the questionnaire, one that produced an especially strong psychological stress due to the repetition of the earthquake in September (questions 13 and 14). Is an intensifying double experience of crisis at the root of the typical reaction of the respondents that was expressed in their answers to the questions proposed in the questionnaire?

The background to the questionnaire. In the discussion of criteria for demarcating the degree of impact, and in the search for indicators of migratory propensity, some quantifiable traits were mentioned, such as the ratio of deaths to total commune population, number of homeless and evacuees out of the commune population, rate of passport applications, tendency to resume telephone communication, proportion of refusals to evacuation when offered, or percentage of people residing in prefabs after their return. Of course, from the outset we had the possibility of questioning the victims themselves, either in the disaster area or in the evacuation zone of the coastal towns, about such indicators of an often dubious predictive power, in order to find out about their will to stay and willingness to rebuild. The difficulties of this procedure, often applied in empirical social research, are plain.

Initially, it was out of the question, because of shortages of funds and personnel in the Department of Geography (Technical University of Munich), to employ a trained interviewer team competent not only in Italian but also in Friulian and its component dialects, and Slovenian in the Resia Valley. Even with the most willing help of the Geography Department in Udine, with which connections were established, the most case-hardened interviewer trying to achieve the daily quota of questionnaires demanded of him would have had to reckon with the fact that a hard-working folk living with trauma and threatened with a potential repetition of the earthquake would not like to be plagued with questions that might simply be too much for them emotionally.[26] Some groups among the people, particularly the older folk, had in some cases no choice except to come back to the unfamiliar crash housing in their old places of residence, which was especially hard to finish because of adverse weather conditions. Possibly some questions intended to test for mobility propensities or willingness to stay could even intensify, with a sort of 'consciousness raising', the awareness of the inescapability of many situations, and weaken the determination to remain active. The social researcher has to let respect for human suffering come before his professional responsibility toward the search for knowledge.

Another very important factor was that a survey was only possible with the permission of the Commissario Straordinario and that any questions asked by regional officials or the Emergency Commissioner could be taken by respondents as a sign of hesitancy about rebuilding, and arouse suspicion that a given village was slated for abolition, and that the written agreement of its former inhabitants was being sought just for that purpose.

On the one hand, the planner needs data in order to locate and adapt the installations necessary for everyday existence properly. On the other hand, he has to take special care to heed the sensibilities of a population stricken by a trauma, even in data gathering, in quite a different fashion than with other decisions. Thus, he has to forego any questions such as those about personal income and financial resources for reconstruction simply because, if nothing else, too-positive answers to them could well lead to the denial of support grants to the respondent.

It was only possible to try to get an approximate idea of the attitude of the people toward remaining and of the next steps they had in mind, in the approaches that are to be recounted. People struck by a catastrophe develop defense mechanisms that can be better understood by applying information and learning theory concepts such as 'reinforcement', 'information storage' and so forth. The cumulative psychological impact of a sequence of earth shocks was reinforced by the shift of epicenters (Fig. 2.4). The series of September quakes occurred around 10 km north of those of the first earthquake in May, and so introduced new population groups to the experience of trauma. The transfer northward toward the Carnic Alps and Karawanken induced a state of alarm in the population of Austria's border province of Carinthia. The consequences of repeated exposure would be worth further examination.

There were two alternative ways to proceed with what follows.

(a) A survey in the evacuation centers, or
(b) a survey in the disaster area.

For extraneous reasons the second course was selected.

In both cases, a very brief written survey with a standardized questionnaire format could be conducted only when an optimal situation for response was technically possible. An oral survey in the disaster area appeared to be too difficult, but there was at first a very large, though constantly decreasing, number of earthquake victims present in the hotel towns on the Adriatic.

The evacuee group, however, displayed certain selective characteristics, in

Figure 2.4 Epicenter distribution, determined by IPG teleseismic system.

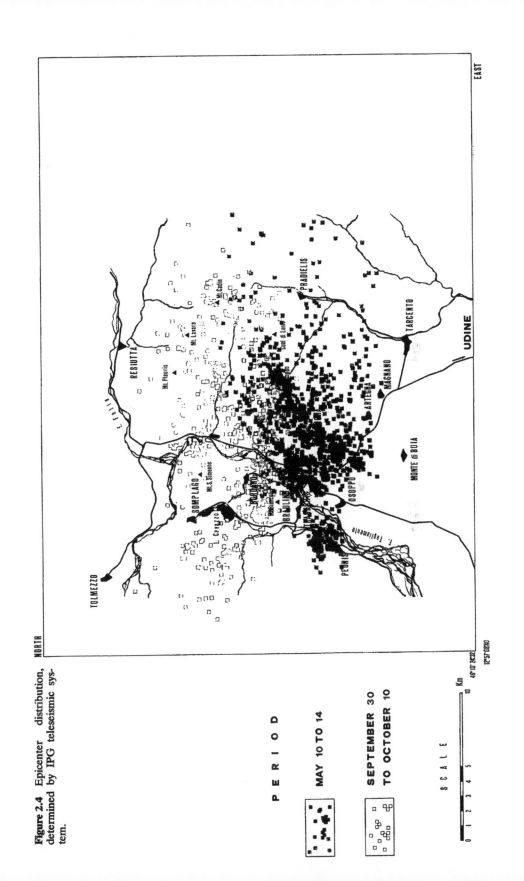

comparison to the original affected population as a whole, that had to be taken into consideration with all findings.

(a) This group was diminished by the number of people in the order of 20 000, who temporarily had left Friuli.

(b) It did not include, of course, any of those individuals who had not left the disaster area either because they had not suffered too severely or, despite earthquake damage, could still manage to find some kind of shelter in the earthquake zone.

(c) Further, there were those who had indeed left the disaster area, but due to personal connections were housed elsewhere than in the official reception areas, for instance with friends in Udine.

(d) Still others had already moved back into completed emergency housing in their home communes. At any rate, the total of evacuees between the high point of reception on October 22, 1976 (32 340 officially reported) and the last day of 1976 (25 383 by the official count) decreased, in the coastal cities and Ravascletto to the tune of 6957 persons. When a survey was planned for February 1977 in the coastal towns, there were still about 20 000 of these people there.

The supposition that the people who had already abandoned the evacuation area represented the more active individuals seemed reasonable, but was not entirely borne out because people were also carrying on their jobs quite actively within the business context of the coastal towns.

The intended respondents of the survey in the evacuation towns therefore constituted only a fraction of a larger whole, whose precise characteristics could not be described and which was not even homogeneous itself with respect to certain features.

The standardized questionnaire reproduced in the Appendix on|p. 195 was designed to be restricted to one page, to be as simple as possible to answer (preferably with check-marks only), and to be processed easily, that is, to permit direct transfer of entries to the data processing unit once the respondent's home address was coded. Beginning with personal data (questions 1–4) and degree of impact (5–13), it went on to inquire about attitudes toward reconstruction (14–15) and how much inclination to mobility might be expected, as well as urban *vs.* rural preferences (16–17). Two supplements, one for the main breadwinner or presumptive decision-maker of the household, and the other for the social workers on the spot, were attached. Translation into Italian and verification of the feasibility of the project from the standpoint of local policy and organizational resources were carried out by Consul Gianfranco Facco Bonetti of the Commissario Straordinario's staff in Udine.

Uncertainty over whether the survey could take place at all dictated that costs be kept to a minimum. The attempt had to be made to get along without outside funds, and without the delay associated with a formal grant (e.g. VW Foundation or DFG, etc.). Between the first discussions with the Emergency Commissioner on December 17, 1976 in Udine and preparation of an interim report outlining the aims of the survey to him, 50 days went by; there would have still been about 60 more days available before the expected end of the evacuation operation on March 31, 1977.

48

The first proposed procedure involved simply distributing (and recovering) the 20 000 questionnaires among all the apartment houses and hotels of coastal towns, with the aid of local hotel staff, apartment managers, social workers etc., to get the data, then collecting it all together in the field headquarters there and bringing everything back to the Central Emergency Office in Udine.

The joint program of work agreed on between the Department of Geography and the State Office for Data Processing led to another field trip on February 6–8, 1977, which served to strengthen ties with the Department of Geography at the University of Udine and permitted an inspection tour of the most perplexing reception commune of Lignano where, on February 7, 12 751 people still lived whose return to their old communes had been stalled.

The accompanying map (Fig. 2.5) shows just where in Lignano the evacuees from particular towns and villages were concentrated. Living so close together meant close information contacts and gave the evacuees an opportunity, in the overabundance of free time they had, to talk about their personal plight. This situation seemed to offer an ideal way of making the distribution and recovery of the questionnaires easier.[27]

A briefing was held for local field agents and their chief assistants by Vice-Prefect Dr Toscano and the author in the operations' center of the Commissario Straordinario in Udine on February 7. Those present considered the questionnaire feasible and foresaw few distribution problems, but some with later recovery, especially in the big reception areas of Lignano and Grado. The respective field agents for each commune received the materials intended for it in sufficient quantities and some in reserve. The investigation seemed to be off to a good start.

Although the questionnaire had been submitted to the Commissario Straordinario for approval and received official assent in a decree of January 25, 1977, the translation into Italian even having been made by the staff of the Emergency Commissioner himself, a freeze on the next phase of the survey operation was imposed that same night for a period of about two months, because of changed developments. The social workers responsible for handing out and collecting the questionnaires were advised, in a strongly worded letter from the Commissario Straordinario on the morning of February 8, that the people then in the evacuation area were being very hesitant about accepting the prefabs set up in their home communes and were in some cases refusing to leave the coastal towns. The social workers would have to concentrate all their efforts on discovering and talking round those who were unwilling to return. Because of this, the survey would also have to be postponed until they had moved into their new prefabs. The forms were brought back to the Prefecture in Udine.

Figure 1.4 on p. 5, a graph of the evacuation operation, shows that the slope of the curve representing returns actually becomes steeper after February 7, so that the measures instituted then were both necessary and effective in the view of the Italian administration. Certainly, the cancellation of plans for the survey at the last minute at first produced bitter disappointment among the research group. The only thing to do was try to make the survey again in less touchy circumstances. For the Emergency Commissioner, the answers to question 15 were particularly crucial: 'What initiative do you

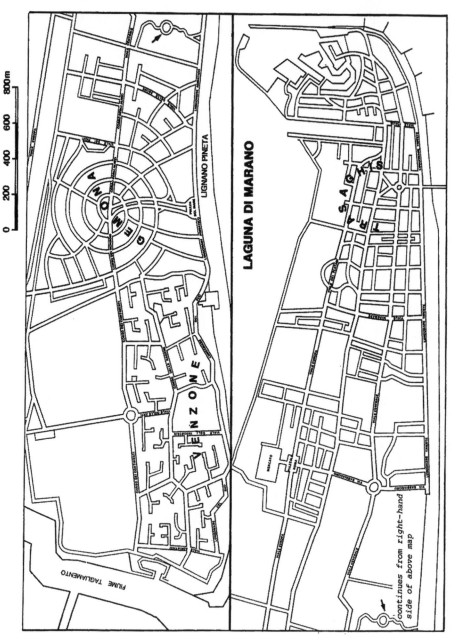

Figure 2.5 Main residential areas of the refugees in Lignano.

intend to take in the near future to improve your present situation?' Positive answers to parts 2, 3 and 4 of question 15, relating to thoughts of emigration, would in some cases have signified rejection of the whole crash building program already under way. But if a survey form itself, amounting to a leading question in such a situation, presents the danger of influencing the subject of inquiry (here, the willingness to return), its use is out of the question.

This offers a parallel, although in a much more realistic form, to research by J. E. Haas and D. S. Mileti on the impact of earthquake predictions. These authors had originally planned to carry out their research according to the Delphi method in two districts – the southern San Francisco Bay Area and San Bernardino County. The earthquake 'prediction', published by CalTech seismologist James H. Whitcomb for the 'Palmdale Bulge' north of Los Angeles on April 20, 1976, forced the research team to confine the investigation only to the Bay Area. Even here, the people questioned might have perceived some connection between the survey and the fact of Whitcomb's 'prediction'.

April and May 1977 were considered a suitable time to try the survey in Friuli again. The subjects to be questioned would be the group who had moved out of the winter evacuation areas into the crash housing quarters. Like the evacuees, these people presented a clearly delimited population for investigation. There would be the advantage that answers would take on a stronger semblance of reality: these people were already living in the prefabs and were back at home in their communes.[28]

Restricting the survey to occupants of the largely segregated prefab quarters offered a similar advantage to a survey in the apartments in the towns on the Adriatic – distributing and recovering the forms could be kept in hand. Therefore, the questionnaire was once more checked over in a sort of emergency session to see if it would still be usable under the changed circumstances, i.e. administering it in the home commune and not the evacuation area. The answer was yes. The research group received assurances again of full official support.

In looking for reasons for the refusal to return home that at the time looked crucial, the close communication contacts that arose from the intimate proximity of former village people living almost one on top of another and condemned to inactivity must be remembered. These communication networks received reports and rumours daily about conditions in the home commune, and how construction of the prefabs was proceeding there. The locations of the future baraccopolis settlements were heatedly discussed on these occasions. The people not working also saw things on occasional bus trips and reported their impressions of what was going on in the communes to those who had stayed at home. Certain key opinion-makers were instrumental in creating varying reactions in this way in different places. The political parties, through their own agencies, participated intensively in the process of opinion-forming. Representatives from Val Belice warned against trusting government measures. The disinclination to return could have rested on highly varied motives, and those responsible for all these operations were faced with the following possibilities.

(a) Did the assessment of risks in locations chosen for the prefab areas in certain particularly exposed communes lying under steep cliffs become a trauma that impaired the inclination to return?

(b)   Or was it the simple fact of prefab structures?

(c)   Did the people try to postpone returning until the last possible moment because of the very bad weather?

(d)   Was there an anxiety about being left to one's own devices again after a period of being taken care of by the government?

(e)   Did getting used to the high standard of vacation apartments built for a demanding leisure market have a kind of acculturation effect and open people's eyes, so that going back to the spartan conditions of the prefabs must have looked like a step backward to them (i.e. comfort *vs.* home attachment, Fig. 2.6)?

(f)   Did some people figure on getting more adequate accommodation by waiting longer and did such hesitation reflect the fact that housing planners, furnishers and manufacturers, as well as the people actually in charge of the work were still in the process of learning, and the first houses were really the most spartan ones?

(g)   Did the evacuated population really prefer to move into the new houses as a body, as a sort of symbolic gesture so that social connections built

Figure 2.6 When it became apparent after the second earthquake of September 15, 1976 that many people would not be able to stay in their home communes during winter, hotels and condominiums in the seaside resorts of the Adriatic (Grado, Lignano, Bibione, Caorle and Jesolo) were commandeered for 32 000 evacuees by the Emergency Commissioner, for an eight-month period. For the former inhabitants of small mountain villages, the transfer into the luxurious apartments of a leisure-class society became a kind of cultural shock. But the rooms, normally used in the summer season, often had no heating, and elderly evacuees and children, particularly, fell ill and suffered depression.

up in the exile areas would not be disrupted, as against the Commissario Straordinario's wish for a smooth flow out of the evacuation areas, so that daily progress could be demonstrated?

(h) Did the people use refusal to accept the prefabs as a way of showing their dissatisfaction with the whole handling of the reconstruction problem?

(i) Was there a fear that accepting the prefabs would be taken to mean that people had resigned themselves to the possibility that the 'temporary' quarters would become permanent, as had happened after the Val Belice earthquake in Sicily?

(j) Did certain political elements regard delaying tactics as a way of putting pressure on the government, knowing that they could exert such pressure via the vulnerable Adriatic tourist industry only as long as the evacuees' going home did not break down the firmness of their stand ('divide et impera')?

(k) Were there some groups that wanted to use the occupied apartments as pawns to put solidarity with other countries in jeopardy, and use the protests of foreign owners dispossessed of their Italian apartments to get a foothold for themselves politically in issues on the international level?

It is not the job of this research to substantiate or refute such conjectures. They are only brought in to show why the original questionnaire in such a confusing situation had to be given up in the evacuation area.

Data collection. The survey described on p. 45, originally intended for the evacuation area, was carried out between April 20 and May 20, 1977 in the home communes. It was directed to heads of households occupying prefabs in the 41 communes declared 'comuni disastrati' by the decree of May 18, 1976 (see Fig. 1.15). At the request of the Prefect of Pordenone Province, the communes therein (Arba, Maniago and Spilimbergo) were included in the survey operation, but they were later excluded from the study.

Official statistics of Udine and Pordenone Provinces on May 1, 1977 show a total of 65 438 persons living in prefabs within the 95 communes to which the crash housing program was applied (Fig. 2.7). Of these:

26 737 (41 per cent) were housed under the Region's plans,
31 741 (48 per cent) were housed by the Commissario Straordinario, and
 7 360 (11 per cent) were housed privately and by relief organizations.

In the 41 communes counting as 'comuni disastrati' by the defining criteria in effect until April 22 when the survey was conducted, 54 867 people lived in prefabs. The questionnaire thus was administered to 83 per cent of the total of Friulians living in prefabs.

Since 20 880 living units had initially been planned, of which 18 279 (or 87.5 per cent) of the whole number were supposed to be erected in the home communes of our survey, it was expected that an equally high number of questionnaires would make up the total. But it became apparent that in the hardest hit and largest commune, Gemona, only occupants of the segregated prefab areas could be approached, but not those who had put up prefabs on their own old private sites scattered all over the town area. So the volume of

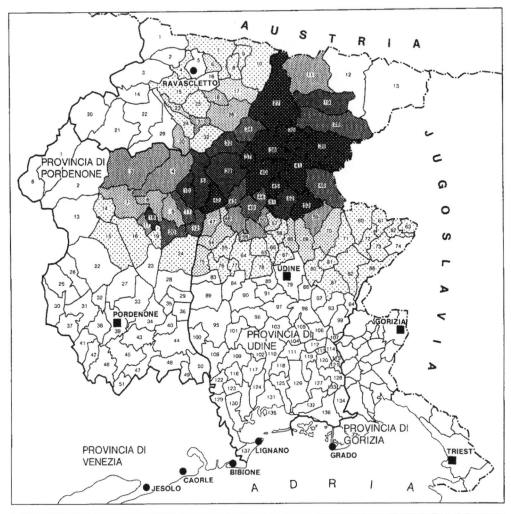

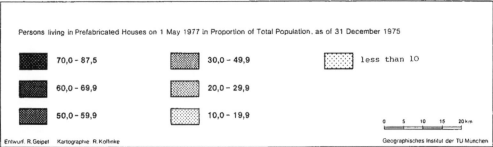

Figure 2.7   Resident population in prefabricated houses as of May, 1 1977.

questionnaires distributed in Gemona was reduced from the planned 2705 to
1056. In the end, 16 630 questionnaires had been handed out in the 'comuni
disastrati' as defined at the time.

In all communes except Gemona, the questionnaires were received by the
communal authority and, after a rather long discussion with the sindaco

54

(mayor) and segretario, were taken out to the sometimes remote subsections of the communes by clerical personnel, social workers, high-school students, boy scouts and other assistants and, after varying periods of time, in the same way collected again. Other communes set up their election ballot boxes into which the sealed envelopes could be dropped.

The rate of response[29] was higher on average in small communes with a modest population size and integrated social networks than in big communes with many subsections and thus much greater difficulties of reassembly. In some communes of the Comunita Montana delle Valli del Torre, such as Attimis and Taipana, Tarcento, Magnano di Riviera, Nimis, Faedis and Lusevera, the inquiry was running in competition with another survey, which led to lower response ratios only in Tarcento. In the problem commune of Gemona, the survey was carried out by Dr G. Meneghel and five assistants and students from the University in Udine, who succeeded in distributing more than 1000 questionnaires in the various prefab areas and getting nearly 500 back after about a week. The rate of return of 46.6 per cent that was achieved there compares with only 18.6 per cent overall when the dispersed prefabs were also counted.

In the process of collecting the forms, whole rows of houses were discovered to be empty in some prefab areas of Gemona, even with repeated visits at different times of day. It was evident, according to the interviewers, that some prefabs had been applied for 'just in case'. Investigation revealed that the people assigned to them were still living in their more or less damaged houses, or somewhere else, and, anyway, only occasionally went to sleep in their prefabs (especially after tremors or when there was 'earthquake weather' as popularly interpreted). Some people with children still in school elsewhere waited till the school year was over before they cared to make use of the houses. The return rate of the questionnaires in the first round of collection can be read on the adjoining map (Fig. 2.8) as an indicator of the administrative potency and co-operation of the communes.

A higher response level than had been expected, in a number of communes, was explained by local authorities in terms of the population's attitude that '. . . the Germans have helped us so much when we needed it most, and now with this survey are the only people who seem to care about the sort of things the questions are about . . .'. Poor returns were explained by the feeling that '. . . so many people are coming around who want something out of us or whisk away documents, maps and pictures and we never hear from them again . . .', which puts the foregoing interpretation into perspective.

Our inquiry actually covered 39 of the 41 'comuni disastrati'. From them, 6568 usable questionnaires were returned out of 15 688 that had been distributed in those communes. The 39.7 per cent response rate can be regarded as extremely high in a disaster area under such circumstances. Another approach made it possible to check it.

Question 4 'How many relatives, apart from yourself, live here with you?', made it possible to total up the number of people represented by each questionnaire. This figure was 20 538 persons. According to official statistics, there were 52 264 persons living in prefabs in the 39 surveyed communes on May 1, 1977. The inquiry was carried out between April 20 and May 20, 1977. The 20 538 persons about whom the study provides information represent 39.3 per cent of all residents of prefabs. The difference between

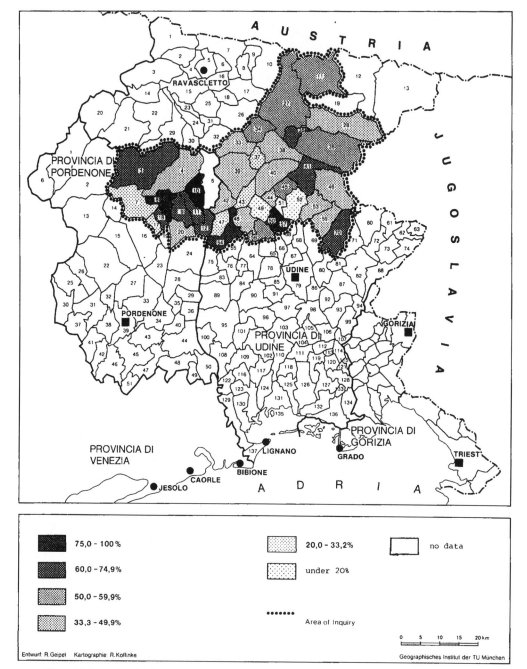

Figure 2.8 Rate of return.

39.3 per cent and 39.7 per cent of a mere 0.4 per cent can possibly be interpreted by slight over-representation of one-person households in the survey, since people living alone, and who were not working, were more likely to be at home. The high average age of 54.06 years for people reporting confirms this surmise.[30]

The answers from 6500 households, representing more than 20 000 persons, give us a certain basis for comparing our findings with those of previous hazard studies. The task of the next section will be to fit what we have found in Friuli into the previous discoveries of other scholars.

2.3 Results from the Friulian questionnaire

The presentation of relationships of the impact of earthquakes to human social systems in the work of White and Haas (1975, p. 322) stops with social and economic disruption. Our task in this section is to make the concept of 'disruption' more operational.

We begin with 'community preparedness', the readiness of a community to cope with a catastrophe, which after all emphatically depends on trends prevailing beforehand. Where these trends were favorable (a youthful population, active, with plenty of job opportunities) the earthquake may even lead to a still more positive economic trend. If there exists a negative situation before the earthquake, on the other hand (an aging or declining population far from jobs), the catastrophe can deal a death blow to the commune (Fig. 2.9). Our observations, therefore, are directed to regions of diverse elements, rather than, as in many instances in the existing literature, to a single stricken city.

It is particularly useful, in the application we propose of other scholars' findings concerning events in Friuli, to follow their schemes and try to fit conclusions from our questionnaire material into them. Therefore, the 'natural events system of the primary hazard' was introduced in Chapter 1 in order that we now can fit in the picture of the 'human community at risk' from our interviews, i.e. our data on the population.

There were 6568 questionnaires available for the evaluation carried out by the Bavarian State Office for Data Processing. The advanced average age of 54.06 years (median 55.513) may not be a matter simply of the aging process of the population of Friuli previously described. The instructions asked the present family head to provide the answers for the whole family. With the traditional Italian family structure, this meant that the oldest family member

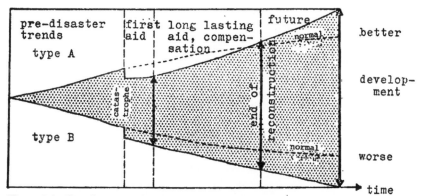

Figure 2.9 Development of active and passive communes.

Table 2.1 Numbers of members of households represented by those interviewed.

No. of persons in household	No. of interviewees	% of total no. of interviewees	Members represented
1	1161	17.7	1161
2	1510	23.0	3020
3	1381	21.1	4149
4	1246	19.0	4984
5	727	11.1	3635
6	316	4.8	1896
7	132	2.0	924
8	55	0.8	440
9	15	0.2	135
10	9	0.1	90
11	5	0.1	55
12	3	0.0	36
13	1	0.0	13
missing value	5	0.1	
total	6568	100.0	20 538

belonging to the household assumed this task when the actual breadwinner was away at work. The respondents included 4045 men (61.6 per cent) and 2403 women (36.6 per cent); this item was unanswered in 120 cases (1.8 per cent). Question 4, 'How many relatives, apart from yourself, live here with you?', made it possible to determine how many people altogether were covered by each response (Table 2.1).

Differences in family structures among communes produce a splitting off of positive and negative axes within our enlarged model.

The large proportion of 'unemployed' (Table 2.3) has to be understood in the light of the high average age of respondents, that is, taking account of the fact that many of them were already living on retirement income or pensions. The proportion of only 5.9 per cent absentees with jobs seems not specially high in view of the 3.3 per cent reported for the whole region in the census of 1971. About a fifth of those questioned, in any case, had changed their place

Table 2.2 Question 9: 'Do you own land in your commune of residence?'

no reply	154	2.3%
yes	4371	66.5%
no	2043	31.1%
total	6568	100.0%

Table 2.3 Question 10: 'Is the main wage earner of your family at present in employment?'

no reply	888	13.5%
yes, in Friuli	3451	52.5%
yes, in Italy	119	1.8%
yes, abroad	268	4.1%
no	1841	28.0%
missing value	1	0.0%
total	6568	100.0%

Table 2.4 Question 11: 'Are you in the same place of work as before the earthquake?'

no reply	1817	27.7%
yes	3395	51.7%
no	1356	20.6%
total	6568	100.0%

of employment since the earthquake (Table 2.4). The high figure for 'no reply' results from many pensioners not considering the question to apply to them. Since, according to the age table from question 2, 40 per cent of the respondents were over 60 years old and 30 per cent over 65, there are also such high figures for question 12.

The stress in our sample on older people is shown clearly in Table 2.5 which indicates that 42.3 per cent of the interviewees were older than 60, while the population of the 39 communes under survey had only 24.4 per cent of this age. This is also shown by the comparative figures for the census of October 24, 1971, given in Figure 2.10. This unfavourable age pattern in many communes makes the success of reconstruction efforts in them doubtful.

Table 2.8 presents the typical picture of a still only slightly urbanized region with most residences in the form of single-family homes. It is notable that the 'no reply' category is the lowest here of anywhere in the whole study.

In studying mobility, it is important to look into a decision-maker's bonds in the form of real estate in the home commune. In the present case, two-thirds of the respondents proved to be landholders (Table 2.9). Subsequent tabulations will show that home- and landownership like this contribute strongly to stability.

Question 12 (Table 2.10) was not only relevant to the respondents but could apply to younger interviewees too if they lived in large households with older relatives.

The responses to question 14 (in Table 2.12) shows that the people who wanted to rebuild were more frustrated by the beginning of winter than by the initial events, and that 70 per cent had not yet begun to rebuild. The next

Table 2.5 Variations in the proportions of those above 60 years old.

Age group above 60	Number	Total	%
Udine Province	103 789 out of	516 910	20.1
Pordenone Province	47 445	253 906	18.7
in the 28 communes of Udine Province surveyed	20 124	85 129	23.6
in the 11 communes of Pordenone Province surveyed	4 194	14 541	28.8
in all the 39 communes	24 318	99 660	24.4
among those interviewed from the 28 communes of Udine Province	2 217	5 435	40.8
among those interviewed from the 11 communes of Pordenone Province	558	1 133	49.3
among all persons interviewed	2 775	6 568	42.3

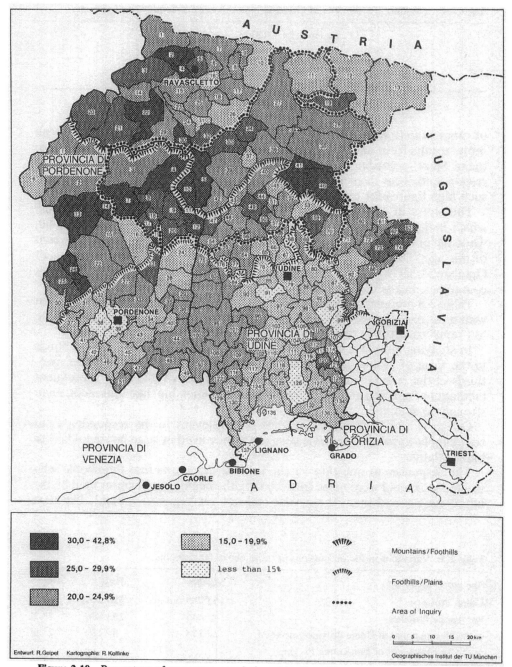

Figure 2.10 Percentage of commune population older than 60 years (census 1971).

question will reveal that a large group of people affected is either not ready or not able to rebuild in the future (Table 2.13). This question, which was perhaps the most important, showed that at the time of the study the determination to rebuild (40.9 per cent) and an attitude of resignation (43.1 per cent), as far as our questions could get at them, were almost evenly balanced

Table 2.6 Question 5: 'Did you lose any near relatives as a result of the earthquake?'

no reply	203	3.1%
yes	788	12.0%
no	5577	84.9%
total	6568	100.0%

Table 2.7 Question 6: 'Did you lose your job as a result of the earthquake?'

no reply	346	5.3%
yes	741	11.3%
no	5481	83.5%
total	6568	100.0%

Table 2.8 Question 7: 'Did you own your house before the earthquake?'

no reply	97	1.5%
yes	5080	77.3%
no	1391	21.2%
total	6568	100.0%

Table 2.9 Question 8: 'Did you own a non-agricultural business before?'

no reply	309	4.7%
yes	855	13.0%
no	5404	82.3%
total	6568	100.0%

Table 2.10 Question 12: 'Is any member of your family receiving a pension?'

no reply	324	4.9%
yes	4263	64.9%
no	1981	30.2%
total	6568	100.0%

Table 2.11 Question 13: 'Which earthquake destroyed your previous apartment or house?'

no reply	316	4.8%
only the May earthquake	2248	34.4%
only the September earthquake	242	3.7%
both earthquakes	3246	49.4%
none of these	516	7.9%
total	6568	100.0%

Table 2.12 Question 14: 'Has some member of your family already taken the initiative for the reconstruction or repair of your old dwelling?'

no reply	569	8.7%
yes, after the May earthquake	878	13.4%
yes, after the September earthquake	173	2.6%
yes, after both earthquakes	409	6.2%
no	4539	69.1%
total	6568	100.0%

Table 2.13 Question 15: 'What initiative do you intend to take in the near future to improve your present situation?'

no reply	695	10.6%
begin reconstruction	2685	40.9%
move to another part of Friuli	54	0.8%
leave Friuli permanently	11	0.2%
move away temporarily and return later	37	0.6%
remain in the emergency building	2833	43.1%
start reconstruction out of the emergency building	253	3.9%
total	6568	100.0%

among the respondents. The number of people (102, or 1.6 per cent) who wished to leave Friuli either for a time or for good was remarkably small, which may, of course, be due simply to the ages of the people questioned, among whom many could no longer take these alternatives. The 253 persons (3.9 per cent) who checked off both of the mutually exclusive categories of starting to rebuild and remaining in the emergency accommodations were acting in a cautious non-commital way, maybe to indicate how long they thought it would be until reconstruction. Later correlations will show which particular groups these were.

Question 16 (Table 2.14) was to test what was learnt from the earthquake.

Since the traditional building style in Friuli uses stone, the 'defection' of about a sixth of the respondents signified a change in outlook. It may have reflected the fact of living in a wooden house at the time and having to remain there quite a while. Some of the notes accompanying questionnaires were outspokenly positive about wood construction, however, so that something more than adaptation to the inevitable has to be assumed.

Although question 15 revealed a propensity to move among only 102 people (1.6 per cent), slightly more of the respondents could conceive of themselves as living ever after in a city (Table 2.15). Of course, rural people normally would say 'no', but many of them had migrated to and lived in big

Table 2.14 Question 16: 'In what type of house would you prefer to live in the future?'

no reply	347	5.3%
in a wooden house resistant to earthquake	1110	16.9%
in a cement/brick dwelling resistant to earthquake	5111	77.8%
total	6568	100.0%

Table 2.15 Question 17: 'Could you imagine yourself living for the rest of your life in a town like Udine or Trieste?'

no reply	261	4.0%
yes	610	9.3%
no	5697	86.7%
total	6568	100.0%

cities before, and especially the younger ones would have been supposed to agree to living in a city. It ought to be noted in passing that it would have been better to name some other, preferably smaller cities along with Udine and to leave Trieste out completely, because there is too much feeling in Friuli against the regional capital.

Thirty-two per cent of the questionnaires only contained information about the prefab-type houses the respondents lived in. As was the case with other indicators of the efficiency of commune administration, here too the communes of Pordenone Province (41.1 per cent of which gave information on the house type) were ahead of those under Udine, which only came to 30.1 per cent. The journal *Ricostruire* carried a table in its first issue showing prefabs erected in Friuli according to their type and place of construction. It revealed only 37 kinds. The respondents, however, named 49 types altogether, sometimes with names they invented themselves. For example, they called the prefabs built by the Haas-Fellbach firm in Baden-Wurttemberg 'Stoccarda' or Stuttgart. This type was largely a gift of the World Lutheran Fellowship.

Despite the fact that the kinds of houses put up differed regionally and in many communes there was no information about them, and that the percentage values are not fully comparable because 335 of those named were not in the table, it was possible to use frequencies of questionnaire response to check on the validity of what was being learnt from other independent data.

Leaving out the 353 answers to the house-type question, for which Table 2.16 does not give percentage values, and letting the 1749 answers that remain out of the total of 2102 equal 100 per cent, we get the ranking shown in Table 2.16. This comparison, covering only totals of 50 or more – which

Table 2.16 Comparison of the prefabricated houses.

Type	Code no.	No. in survey	Survey (%)	List (%)
Volani	1	393	22.46	13.72
Valentina	2	279	15.95	10.92
Krivaja	3	259	14.80	8.42
Pittini	4	101	5.77	5.87
Tecna	5	84	4.80	5.46
Atco	6	168	9.60	5.10
Cocel	10	87	4.97	3.41
Carniche	14	62	3.54	2.50
Meccanocar	17	60	3.43	1.94
Ceccoli	25	51	2.91	0.89
10 types (= 88.72%)		1544		

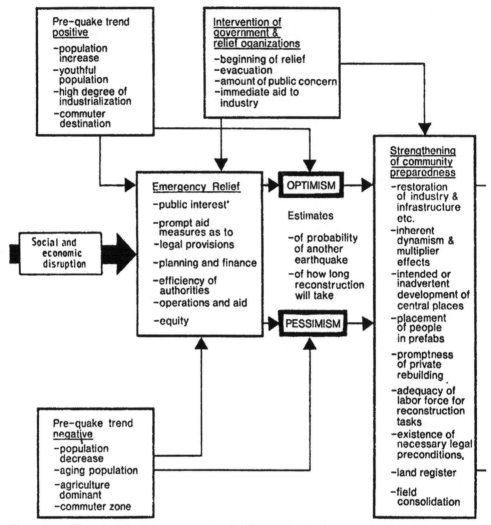

Figure 2.11 The principles that govern regional differentiation in the reconstruction process.

together made up 88 per cent of the whole – greatly over-represents the Atco type only. Otherwise, the ordering of prefabs from the questionnaire corresponds approximately to the ranking found in official statistics.

The model shown in Figure 2.11 depicts what we have learned from Friuli. It is based on 'Social and economic disruption', referred to by White and Haas (cf. Fig. 2.3), and looks into the principles that govern regional differentiation in the reconstruction process.

2.4 The social circumstances of the respondents of our survey one year after the disaster

Just a year after the catastrophe, the following situation had been reached. The evacuees had come back from the Adriatic towns, moved into the prefabs

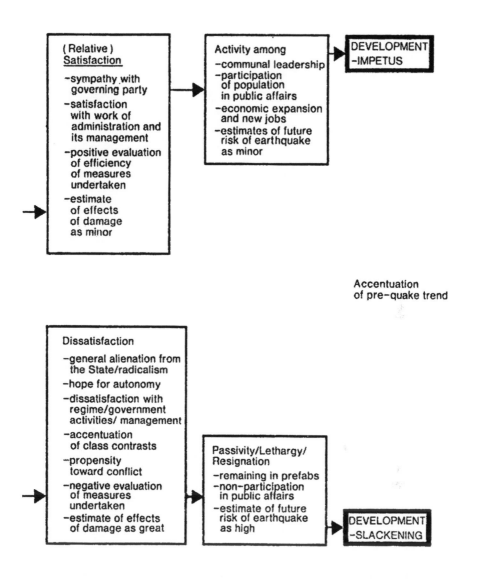

(Relative) Satisfaction	Activity among	DEVELOPMENT -IMPETUS
−sympathy with governing party −satisfaction with work of administration and its management −positive evaluation of efficiency of measures undertaken −estimate of effects of damage as minor	−communal leadership −participation of population in public affairs −economic expansion and new jobs −estimates of future risk of earthquake as minor	

Accentuation
of pre-quake trend

Dissatisfaction	Passivity/Lethargy/ Resignation	DEVELOPMENT -SLACKENING
−general alienation from the State/radicalism −hope for autonomy −dissatisfaction with regime/government activities/ management −accentuation of class contrasts −propensity toward conflict −negative evaluation of measures undertaken −estimate of effects of damage as great	−remaining in prefabs −non-participation in public affairs −estimate of future risk of earthquake as high	

and were now waiting for the permanent reconstruction plans for their towns and cities (Fig. 2.12). At this phase, the people we interviewed fell into two groups of nearly equal size: one being those inclined to proceed immediately with reconstruction, the others at first hesitating, wanting to stay in the pre-fabs and perhaps, because of age, illness or insufficient resources, destined to remain for good. This is probably the time that is hardest for people to bear, after every catastrophe.

The heroic efforts to rescue dead and injured were past, as was the sojourn in a new and unaccustomed evacuation environment. There were no more headlines about Friuli in foreign papers, no TV news teams betrayed the participation of sensation-seeking strangers in their private misery. Relief shipments from abroad were getting sparser or becoming exhausted, and the public throughout the world had long since turned away to new sensations.

What occurs in such a period of indecision? What impulses and influences

Figure 2.12 After the May earthquake, the population was housed provisionally in tents. A long-lasting period of rain set in immediately after the earthquake and turned life in the tent cities into a nightmare. The motto 'Dalle tende alle case' ('from the tents straight back to the houses') could not be accomplished. After the evacuation into the coastal towns, the population had to return to the prefabs.

are projected from the past on to a mournful present and an unknown future? The return to normality involves clearing many bureaucratic hurdles. Application forms must be filled out, losses must be proven, extracts made from the land registers, and individual financial means for reconstruction must be critically assessed. The solidarity that arose under the pressure of such need is relaxing again. To be sure, the rich and poor alike are living side-by-side in barracks, but some people already stand out by virtue of superior connections with officialdom, construction firms, or legal advisers. Men who are employed in growth industries with great influence not only earn more but are better connected with sources of information than those who only work in the fields or simply live on pensions. The more optimistic, energetic, active people already begin to climb up out of the general misery in the barrack town. The people with more imagination, enterprising spirit, a poverty of scruples, come to understand the opportunities of a fresh start.

The social geographer, of course, is especially interested in the spatial aspect – in the way his results vary from commune to commune. This will be shown in Section 3.3. Right now there are some questions that have already been posed in the Preface about more general and typical behavior patterns which ought to be answered from the survey data. This is done in the ensuing section by the interpretation of two- and three-dimensional matrices designed to illustrate relationships among three complexes:

(a) demography and social situation;
(b) impact of the earthquake;
(c) reaction to the catastrophe.

How do the demographic and social situation, with the impact of the earthquake, together affect the way the population reacts?

The demographic and social situation can be determined from the answers

66

respondents made concerning such items as age, family size, home ownership, property holdings within the commune, and the breadwinner's place of employment. Impact of the earthquake is represented by the loss of relatives, loss of work place, and having to live in a certain type of prefab. The type of reaction can be seen from what was said about repairs undertaken in May and September, willingness to rebuild, readiness to live in a city, and readiness to live in a wooden house.

In what follows, the relationships among these three complexes are first summarized in a diagram (Fig. 2.13). The most notable relationships will then be examined with the aid of tables.

The introductory diagram shows that significant relationships of varying

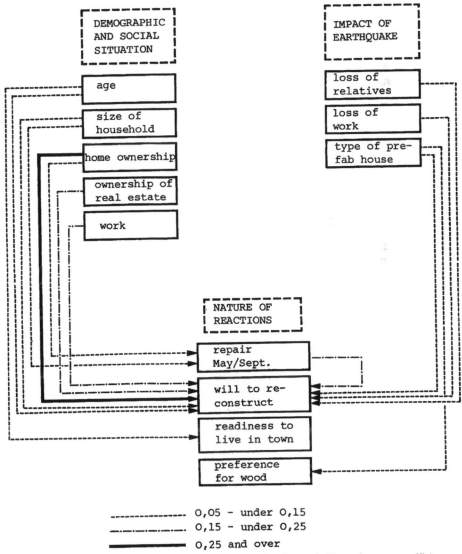

Figure 2.13 Significant relationships according to the Cramer's V contingency coefficient.

intensities exist between the causal complexes and the types of reaction. We used Cramer's V to quantify them. This is a correlation coefficient derived from the chi-square test that measures the degree of relationship between nominally scaled variables. It is designed to be used to show relationships where both nominal and scalar variables are involved. Some tables also give chi-square values, degrees of freedom, significance, contingency coefficients and gamma values. The relationships depicted in the diagram will be examined with the aid of tables in the next sections, and may be interpreted according to the following principles. A high chi-square value indicates a strong relationship between two variables. The table size is represented through the degrees of freedom. The significance level gives the most important measure of relationships. It represents the probability that the chi-square value will occur purely randomly in a distribution with no systematic relationship among the variables. The lower the significance level, the stronger is the relationship, and so the interdependence of the variables. In Table 2.20, for instance, the significance value of 0.000 is given. This means that the probability that the values recorded could occur by chance is less than 1 in 1000. So we can also say that the relationship is significant at the 1 : 1000 level. With a value of 0.0, the significance level is even higher still, for the program stops calculating at the fifth place to the right of the period.

The correlation coefficient, like Cramer's V, is derived from the chi-square test. It is suitable for a comparison of two series of equal length. While so-called coefficients allow inferences about variables arranged nominally as well as scaled numerically, gamma measures the relationships between ordinally scaled variables.

The demographic and social situation

Age. It is worth noticing that the intermediate age cohorts (30–59) show similar values for both alternatives – to begin rebuilding or to remain in the prefabs – while the groups both under 30 and over 60 differ from them (Table 2.17).

The discrepancies among the mean values give the pattern shown in Table 2.18. The younger people are definitely not more 'dynamic'. It is more likely that they are acting tentatively. Furthermore, they are not so much caught up in home ownership, so that it also makes more sense for them to stay in prefabs for this reason (Table 2.19).

The lower dynamism of the younger respondents also finds expression in the fact that, although there is a clear tendency for younger people to be able to imagine themselves living in a city, in this they do not differ from their elders very much (Table 2.20). We shall have more to say later about behavioral differences possibly caused by age. Family size, however, seems to have a greater influence than age.

Family size. Since the questionnaire asked how many relatives the person giving the answers lived with in one household (see Table 2.1), the influence of household size on manner of reaction could be determined. The choice between the alternatives of rebuilding or remaining in emergency accommodation showed the inclination to remain in a prefab as decreasing as family size increased (Table 2.21). With increase in number of family members, the frequency with which it was reported that repairs had already been under-

68

Table 2.17 Cross tabulation of activity with age: question 15 by question 2 (age).

		Under 30	30–45	45–60	60 and older	Row total
	COUNT I ROW PCT I COL PCT I TOT PCT I	1.00I	2.00I	3.00I	4.00I	
reconstruct	1.00 I I I I	185 I 6.9 I 39.9 I 3.2 I	622 I 23.2 I 49.4 I 10.6 I	858 I 32.0 I 50.3 I 14.6 I	1020 I 38.0 I 41.8 I 17.4 I	2685 45.7
remain in prefab	2.00 I I I I	254 I 9.0 I 54.7 I 4.3 I	547 I 19.3 I 43.4 I 9.3 I	746 I 26.3 I 43.7 I 12.7 I	1286 I 45.4 I 52.7 I 21.9 I	2833 48.2
other	3.00 I I I I	25 I 7.0 I 5.4 I 0.4 I	91 I 25.6 I 7.2 I 1.5 I	103 I 29.0 I 6.0 I 1.8 I	136 I 38.3 I 5.6 I 2.3 I	355 6.0
column total		464 7.9	1260 21.5	1707 29.1	2442 41.6	5873 100.0

```
CHI SQUARE =     54.65251 WITH    6 DEGREES OF FREEDOM    SIGNIFICANCE =  0.0000
CRAMER'S V =     0.06821
CONTINGENCY COEFFICIENT =     0.09602
KENDALL'S TAU B =     0.03217  SIGNIFICANCE =    0.0001
KENDALL'S TAU C =     0.02986  SIGNIFICANCE =    0.0003
GAMMA =     0.05178
SOMER'S D =     0.02884

NUMBER OF MISSING OBSERVATIONS =      695
```

taken in May or September grew, because of greater need but also because the number of wage earners might sometimes be larger (Table 2.22).

Home and real estate ownership. Owning one's own house and land within the commune predisposed strongly to going ahead with reconstruction. Thus, 26.7 per cent of home owners as against 14.9 per cent of renters had already begun repairs before being evacuated (Tables 2.23 and 24).

The non-home owners, who dwelt within their 'own four walls' for the first time when a prefab was assigned to them, showed much less inclination to move out of these prefabs when asked about their plans for improving their situation. A problem may arise in this connection with resettlement later on in rental units when rebuilt (Table 2.25).

Table 2.18 Age differentiation in relation to improvement of situation, as a departure from the mean.

	−30	30–45	45–60	Over 60	Mean
reconstruction	−5.8	+4.3	+5.4	−3.9	45.7
remain in prefab	+6.5	−4.8	−4.5	+4.5	48.2

Table 2.19 Age differentiation in relation to house ownership.

	−30	30–45	45–60	Over 60	Mean
house owners	54.25	69.93	83.09	84.07	78.38%

Table 2.20 Cross tabulation of living in a town with age: question 17 by question 2 (age).

		Under 30	30–45	45–60	60 and older	Row total
	COUNT I ROW PCT I COL PCT I TOT PCT I	1.00 I	2.00 I	3.00 I	4.00 I	
would go to city	1.00 I I I I	74 12.1 15.3 1.2	153 25.1 11.3 2.4	185 30.3 10.2 2.9	198 32.5 7.4 3.1	610 9.7
oppose going to city	2.00 I I I I	411 7.2 84.7 6.5	1197 21.0 88.7 19.0	1627 28.6 89.8 25.8	2462 43.2 92.6 39.0	5697 90.3
column total		485 7.7	1350 21.4	1812 28.7	2660 42.2	6307 100.0

```
CHI SQUARE =      37.30487 WITH    3 DEGREES OF FREEDOM    SIGNIFICANCE =   0.0000
CRAMER'S V =      0.07691
CONTINGENCY COEFFICIENT =       0.07668
KENDALL'S TAU B =       0.06836  SIGNIFICANCE =    0.0000
KENDALL'S TAU C =       0.04740  SIGNIFICANCE =    0.0000
GAMMA =      0.19059
SOMER'S D =      0.03445

NUMBER OF MISSING OBSERVATIONS =       261
```

Table 2.21 Cross tabulation of improvement in situation with size of family: question 15 by question 4.

		Number of persons			5 or more	Row total
		1	2	3–4		
	COUNT I ROW PCT I COL PCT I TOT PCT I	1.00 I	2.00 I	3.00 I	4.00 I	
reconstruct	1.00 I I I I	319 11.9 31.8 5.4	561 20.9 41.9 9.6	1169 43.5 48.9 19.9	636 23.7 55.8 10.8	2685 45.7
remain in prefab	2.00 I I I I	625 22.1 62.3 10.6	712 25.1 53.1 12.1	1067 37.7 44.6 18.2	429 15.1 37.6 7.3	2833 48.2
other	3.00 I I I I	59 16.6 5.9 1.0	67 18.9 5.0 1.1	154 43.4 6.4 2.6	75 21.1 6.6 1.3	355 6.0
column total		1003 17.1	1340 22.8	2390 40.7	1140 19.4	5873 100.0

```
CHI SQUARE =     161.91397 WITH    6 DEGREES OF FREEDOM    SIGNIFICANCE =   0.0
CRAMER'S V =      0.11741
CONTINGENCY COEFFICIENT =       0.16380
KENDALL'S TAU B =      -0.12093  SIGNIFICANCE =    0.0
KENDALL'S TAU C =      -0.11427  SIGNIFICANCE =    0.0
GAMMA =     -0.19092
SOMER'S D =     -0.10647

NUMBER OF MISSING OBSERVATIONS =       695
```

Table 2.22 Cross tabulation of start of repair with size of family: question 14 by question 4.

		Number of persons				
		1	2	3–4	5 or more	Row total
	COUNT I ROW PCT I COL PCT I TOT PCT I	1.00I	2.00I	3.00I	4.00I	
repair	1.00 I I I I	165 I 11.3 I 15.6 I 2.8 I	298 I 20.4 I 21.7 I 5.0 I	641 I 43.9 I 26.1 I 10.7 I	356 I 24.4 I 30.4 I 5.9 I	1460 24.3
no repair	2.00 I I I I	828 I 18.2 I 83.4 I 13.8 I	1078 I 23.7 I 78.3 I 18.0 I	1817 I 40.0 I 73.9 I 30.3 I	816 I 18.0 I 69.6 I 13.6 I	4539 75.7
column total		993 16.6	1376 22.9	2458 41.0	1172 19.5	5999 100.0

```
CHI SQUARE =      64.76503 WITH    3 DEGREES OF FREEDOM    SIGNIFICANCE =   0.0000
CRAMER'S V =      0.10390
CONTINGENCY COEFFICIENT =      0.10335
KENDALL'S TAU B =    -0.09471   SIGNIFICANCE =   0.0
KENDALL'S TAU C =    -0.09713   SIGNIFICANCE =   0.0
GAMMA =    -0.18538
SOMER'S D =    -0.06802

NUMBER OF MISSING OBSERVATIONS =      569
```

Table 2.23 Cross tabulation of improvement in situation with house ownership: question 15 by question 7.

		Yes	No	Row total
	COUNT I ROW PCT I COL PCT I TOT PCT I	1.00I	2.00I	
reconstruct	1.00 I I I I	2421 I 91.3 I 53.3 I 41.8 I	231 I 8.7 I 18.4 I 4.0 I	2652 45.8
remain prefab	2.00 I I I I	1842 I 65.9 I 40.5 I 31.8 I	952 I 34.1 I 76.0 I 16.4 I	2794 48.2
other	3.00 I I I I	280 I 80.0 I 6.2 I 4.8 I	70 I 20.0 I 5.6 I 1.2 I	350 6.0
column total		4543 78.4	1253 21.6	5796 100.0

```
CHI SQUARE =      517.07788 WITH    2 DEGREES OF FREEDOM    SIGNIFICANCE =   0.0
CRAMER'S V =      0.29869
CONTINGENCY COEFFICIENT =      0.28619
KENDALL'S TAU B =     0.25356   SIGNIFICANCE =   0.0
KENDALL'S TAU C =     0.21986   SIGNIFICANCE =   0.0
GAMMA =     0.54955
SOMER'S D =     0.32437

NUMBER OF MISSING OBSERVATIONS =      772
```

Table 2.24 Cross tabulation of improvement in situation with real-estate ownership: question 15 by question 9.

		Yes	No	Row total
	COUNT I ROW PCT I COL PCT I TOT PCT I	1.00I	2.00I	
		--------I--------I--------I		
recontruct	1.00 I I I I	2072 I 78.5 I 52.6 I 36.0 I	567 I 21.5 I 31.2 I 9.8 I	2639 45.8
	-I--------I--------I			
remain in prefab	2.00 I I I I	1634 I 58.8 I 41.5 I 28.4 I	1146 I 41.2 I 63.0 I 19.9 I	2780 48.3
	-I--------I--------I			
other	3.00 I I I I	232 I 68.6 I 5.9 I 4.0 I	106 I 31.4 I 5.8 I 1.8 I	338 5.9
	-I--------I--------I			
column total		3938 68.4	1819 31.6	5757 100.0

```
CHI SQUARE =   244.03633 WITH   2 DEGREES OF FREEDOM   SIGNIFICANCE =  0.0
CRAMER'S V =   0.20589
CONTINGENCY COEFFICIENT =     0.20166
KENDALL'S TAU B =    0.17812  SIGNIFICANCE =  0.0
KENDALL'S TAU C =    0.17421  SIGNIFICANCE =  0.0
GAMMA =    0.35282
SOMER'S D =    0.20151

NUMBER OF MISSING OBSERVATIONS =    811
```

It is no wonder that a large proportion (46.9 per cent) of the elderly among the former house owners wanted to stay in the prefabs, for they did not have time left to think of reconstruction. But the younger home owners themselves were also hesitant, which may have had quite a bit to do with their tendency to emigrate if necessary.

If we look now under 'other solutions' in Table 2.26 to assess the intention to improve the situation, by beginning to rebuild as soon as possible, the group of people who already had a house before was naturally also motivated.

Table 2.25 Age differentiation according to the will to remain in the prefabricated buildings.

	Under 30	30–45	45–60	Over 60
those wishing to remain in the prefab buildings				
house owners (1842)	39.8 (99)	32.8 (286)	36.3 (508)	46.9 (949)
non-house owners (952)	72.9 (153)	67.7 (254)	79.3 (226)	83.3 (319)
those preferring other solutions				
house owners (2701)	60.2 (150)	67.2 (586)	63.7 (892)	53.1 (1073)
non-house owners (301)	27.1 (57)	32.3 (121)	20.7 (59)	16.7 (64)
corrected chi-square	49.07	128.93	176.44	169.29
significance	0.0000	0.0	0.0	0.0
contingency coefficient	0.31	0.30	0.30	0.25
gamma	−0.60	−0.62	−0.74	−0.69
$N = 5796$				

Table 2.26 Age differentiation according to the will to reconstruct.

	Under 30	30–45	45–60	Over 60
those wishing to start with the reconstruction				
house owners (2421)	53.8 (134)	60.6 (528)	57.4 (803)	47.3 (956)
non-house owners (231)	22.9 (48)	23.7 (89)	16.1 (46)	12.5 (48)
those preferring other solutions				
house owners (2122)	46.2 (115)	39.4 (344)	42.6 (597)	52.7 (1066)
non-house owners (1022)	77.1 (162)	76.3 (286)	83.9 (239)	87.5 (335)
corrected chi-square	44.34	140.72	159.27	158.44
significance	0.0000	0.0	0.0	0.0
contingency coefficient	0.29	0.31	0.29	0.24
gamma	0.59	0.66	0.75	0.72
$N = 5796$				

Here too, the young were hesitant and yielded values intermediate between those of people over 60 and the middle aged.

The question of how much the response was affected by whether or not the main wage earner of the household had a job can be mentioned as a final causal factor within the demographic and social situation.

Occupation of the main wage earner. Among the four components of the demographic and social situation mentioned so far, home- and landowner-ship had a much greater influence on how people reacted than did family size or even age. The matter of whether or not the main wage earner had a job must be considered in the light of the high average age of the respondents, many of whom were already retired (cf. 42.3 per cent over 60). The factor of the wage earner's employment thus takes on intermediate importance as an influence on how people react (Table 2.27).

Since people who had jobs were readier to reconstruct, particular attention had to be devoted to the maintenance and increase of employment levels. The fact that numerous Friulians undertook the reconstruction of the factories first, and only afterwards of their homes, has been alluded to before. How-ever, other data showed that an influence exerted by the loss of near relatives upon future building activity, or the wish to leave, could not be detected.

Loss of work place. The factor of whether or not the main wage earner had a job at all seemed to have much more to do with the character of the reaction than did the matter of whether or not he worked at the same place as before the earthquake (cf. question 6). This question seemed to have been both quantitatively and qualitatively meaningless; only a mere 639 respondents (11.4 per cent) had had to change their working place, while 4952 (88.6 per cent) did not.

Temporary housing in particular types of prefabs. It was mentioned before that only around 32 per cent of the respondents answered the question that had been added at the last minute about prefab type, and that in the data processing only house types that could be taken as representative because they occurred with frequencies of approximately 100 were considered.

Table 2.27 Cross tabulation of improvement in situation with main wage earner in employment: question 15 by question 10.

		Work	No work	Row total
	COUNT I ROW PCT I COL PCT I TOT PCT I	1.00I	2.00I	
	1.00 I	1813 I	637 I	2450
reconstruct	I	74.0 I	26.0 I	47.3
	I	52.1 I	37.6 I	
	I	35.0 I	12.3 I	
	2.00 I	1455 I	978 I	2433
remain in prefab	I	59.8 I	40.2 I	47.0
	I	41.8 I	57.7 I	
	I	28.1 I	18.9 I	
	3.00 I	214 I	79 I	293
other	I	73.0 I	27.0 I	5.7
	I	6.1 I	4.7 I	
	I	4.1 I	1.5 I	
column total		3482	1694	5176
		67.3	32.7	100.0

```
CHI SQUARE =    116.44711 WITH    2 DEGREES OF FREEDOM    SIGNIFICANCE =  0.0
CRAMER'S V =      0.14999
CONTINGENCY COEFFICIENT =      0.14833
KENDALL'S TAU B =      0.11492  SIGNIFICANCE =    0.0
KENDALL'S TAU C =      0.11330  SIGNIFICANCE =    0.0
GAMMA =      0.22969
SOMER'S D =      0.12865

NUMBER OF MISSING OBSERVATIONS =      1392
```

This question proved highly revealing in view of reported construction scandals and because of the very varied quality of different sorts of prefabs (Fig. 2.14). Among 2039 usable responses concerning house type, there were enough of the Volani model (371), the Valentina (275), the Krivaja (252), the Atco (165) and the Pittini (99) to make a correlation between living in a certain kind of prefab and the nature of a person's reactions.

This gives some grounds for supposing that the willingness to occupy a wooden house in the future is influenced by the experience a person has had with his own prefab (which need not be of wood; Table 2.28). The 'Krivaja' type of wooden house produced in Zavidovici, Yugoslavia, seems to have induced a positive attitude in those who lived in it. Almost 20 per cent of occupants were prepared to change their previous preference in materials, most clearly in the younger age groups (people under 30, 25.9 per cent; 30- to 45-year-olds, 25.5 per cent). The Atco container, a Canadian type, although not itself of wood, affected only 10 of its occupants (2.8 per cent) in this way. Table 2.29 compares the relationships indicated here.

The attitudes of residents toward prefab types, as brought out by the survey, should not be overstressed. A much more comprehensive questionnaire would be required on this point.[31] The correlations given are, on the other hand, something of an accidental result. Here the desire at first to stay in emergency accommodation is not in any clear way influenced by the kind of prefab occupied at the time of the survey: the figures were 46.6 per cent of the residents for the Volani prefabs, 29.0 per cent for the Valentina, 50.2 per

Figure 2.14 The illustration shows two different types of prefabs in juxtaposition. To the left, a wooden type with good insulation; to the right, the Canadian type, 'Atco', furnished with a door and small windows on the front side (cf. Fig. 4.2) and with a bare wall to the rear. The flat sheet-metal roofs were covered at first only with an additional plastic foil. The apportioning of houses of such a striking difference in quality led to tensions between the occupants and the administration.

Table 2.28 Cross tabulation of preference for future house type with type of prefab: question 16 by question 18.

	COUNT I ROW PCT I COL PCT I TOT PCT I	*Volani* 1.00I	*Valentina* 2.00I	*Krivaja* 3.00I	*Pittini* 4.00I	*Atco* 5.00I	*Others* 6.00I	*Row total*
wood	1.00 I I I I	49 I 16.1 I 13.2 I 2.4 I	26 I 8.5 I 9.5 I 1.3 I	50 I 16.4 I 19.8 I 2.5 I	11 I 3.6 I 11.1 I 0.5 I	14 I 4.6 I 8.5 I 0.7 I	155 I 50.8 I 17.7 I 7.6 I	305 15.0
concrete or brick	2.00 I I I I	322 I 18.6 I 86.8 I 15.8 I	249 I 14.4 I 90.5 I 12.2 I	202 I 11.6 I 80.2 I 9.9 I	88 I 5.1 I 88.9 I 4.3 I	151 I 8.7 I 91.5 I 7.4 I	722 I 41.6 I 82.3 I 35.4 I	1734 85.0
column total		371 18.2	275 13.5	252 12.4	99 4.9	165 8.1	877 43.0	2039 100.0

```
CHI SQUARE =      23.83717 WITH    5 DEGREES OF FREEDOM    SIGNIFICANCE =  0.0002
CRAMER'S V =       0.10812
CONTINGENCY COEFFICIENT =       0.10750
KENDALL'S TAU B =    -0.05229  SIGNIFICANCE =   0.0002
KENDALL'S TAU C =    -0.04537  SIGNIFICANCE =   0.0011
GAMMA =    -0.12359
SOMER'S D =   -0.03067

NUMBER OF MISSING OBSERVATIONS =     4529
```

Table 2.29 Age differentiation according to the comparison of 'Krivaja' (best) and 'Atco' (worst) prefabricated building types.

	Under 30	30–45	45–60	Over 60
'Krivaja' (126)	71.4	41.2	53.2	46.3
ϕ	46.6	37.3	38.2	46.0
'Atco' (55)	22.2	23.5	29.3	44.0
difference: 'Krivaja'/'Atco'	49.2	17.7	23.9	2.3

cent for the Krivaja, 51.5 per cent for the Pittini and, for the Atco, 34.6 per cent.

It is understandable that the 'Krivaja' wooden house, suited for permanent use, came off better in the judgment of its residents than the Canadian 'Atco' container. It was ordered on the advice of military authorities because it could be stored in minimal space to constitute a logistic reserve against similar catastrophies after being collected again in Friuli. But the comparison between the prefab types most and least favored, which is rather unacceptable for this reason, reveals something else important – how the different age groups react to having to live in a prefab.

In spite of a poor representation in the individual sections of the corresponding tables, the tendency shown in Table 2.29 can be perceived when a three-dimensional correlation is made with the contrasting pair of 'Krivaja' and 'Atco'.

Whereas the younger people among the prefab occupants were obviously still emphatically aware of differences in quality, and judged and reacted accordingly, resignation and indifference toward particular types of buildings were widespread among the very old (2.3 per cent). The diagram that follows attempted to preserve significant relationships among distinct parameters when they were filtered through age-specific evaluations (Fig. 2.15).

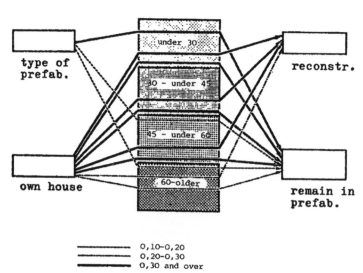

Figure 2.15 Significant correlations according to age and the Cramer's V contingency coefficient.

The earthquake's effect. The earthquake took a toll of human lives, wiped out work places, and destroyed houses and apartments. This last fact need not be elaborated since all respondents were in the same situation, having to live in prefabs. Compared to this factor, other effects produced by the earthquake had an astonishingly minor additional influence on how people reacted. For instance, the will to reconstruct was only slightly affected by whether or not the respondent lost close relatives (Table 2.30).

Repairs made in vain and the will to reconstruct. Reactions such as 'repaired the house in May (or September)', although entered under the rubric of 'character of reaction' also in a way belonged under the rubric of 'impact' just considered, for anyone who started rebuilding in May (that is, 13.4 per cent of the respondents) lost what savings he had put into it in the September earthquake. Nevertheless, the number of people who wished to begin reconstruction in spite of this (64.7 per cent) was half again as great as those who previously delayed repairs, of whom only 41.6 per cent wanted to go ahead with reconstruction (Table 2.31).

The alternative of 'staying in emergency accommodations' was taken into consideration by only 28.2 per cent of the people whose attempts at restoration had once already been in vain. Among the more passive ones, almost twice as many respondents (53 per cent) selected this delaying behavior. The way Friulians behaved during and after the disaster, therefore, is capable of adding more information to the knowledge of how people behave in a disaster situation, described, for example, as relationships among the frequency of

Table 2.30 Cross tabulation of improvement in situation with loss of near relatives: question 15 by question 5.

		Yes	No	Row total
	COUNT I ROW PCT I COL PCT I TOT PCT I	1.00I	2.00I	
reconstruct	1.00 I I I I	269 I 10.2 I 39.2 I 4.7 I	2356 I 89.8 I 46.8 I 41.2 I	2625 45.9
remain in prefab	2.00 I I I I	376 I 13.6 I 54.7 I 6.6 I	2388 I 86.4 I 47.5 I 41.8 I	2764 48.3
other	3.00 I I I I	42 I 12.7 I 6.1 I 0.7 I	288 I 87.3 I 5.7 I 5.0 I	330 5.8
column total		687 12.0	5032 88.0	5719 100.0

```
CHI SQUARE =      14.51448 WITH   2 DEGREES OF FREEDOM   SIGNIFICANCE =  0.0007
CRAMER'S V =      0.05038
CONTINGENCY COEFFICIENT =      0.05031
KENDALL'S TAU B =    -0.04598   SIGNIFICANCE =   0.0000
KENDALL'S TAU C =    -0.03143   SIGNIFICANCE =   0.0002
GAMMA =    -0.13431
SOMER'S D =    -0.07433

NUMBER OF MISSING OBSERVATIONS =      849
```

Table 2.31 Cross tabulation of improvement in situation with start of repair in May/September: question 15 by question 14.

		Repair	No repair	Row total
	COUNT I ROW PCT I COL PCT I TOT PCT I	1.00 I	2.00 I	
reconstruct	1.00 I I I I	815 I 31.8 I 64.7 I 14.9 I	1748 I 68.2 I 41.6 I 32.0 I	2563 46.9
remain in prefab	2.00 I I I I	355 I 13.7 I 28.2 I 6.5 I	2229 I 86.3 I 53.0 I 40.8 I	2584 47.3
other	3.00 I I I I	90 I 28.3 I 7.1 I 1.6 I	228 I 71.7 I 5.4 I 4.2 I	318 5.8
column total		1260 23.1	4205 76.9	5465 100.0

```
CHI SQUARE =    241.81920 WITH    2 DEGREES OF FREEDOM    SIGNIFICANCE =  0.0
CRAMER'S V =     0.21035
CONTINGENCY COEFFICIENT =    0.20585
KENDALL'S TAU B =    0.16703  SIGNIFICANCE =   0.0
KENDALL'S TAU C =    0.14798  SIGNIFICANCE =   0.0
GAMMA =    0.36087
SOMER'S D =    0.20854

NUMBER OF MISSING OBSERVATIONS =    1103
```

aftershocks, property losses and 'anxiety', i.e. basic dispositions such as optimism *vs.* pessimism. It remains to be shown whether this astonishingly high will to resist of large numbers of the stricken Friulian populace can be kept up in the face of longer-term seismic threats, such as became evident in the 6.0 R earthquake of September 17, 1977 and the 4.8 R earthquake of April 18, 1979.

Nonetheless, a few general conclusions that can be built into a gradually expanding hazard theory probably can be drawn from the factual material presented thus far.

The victim of a catastrophe recovers fastest from an accident if he lives in a family, meaning that he shares responsibility for other family members, and if he has kept his job. Landholding – that is, attachment to one's house and ground – is a powerful motive for reconstruction. Younger persons, on the contrary, who have no household of their own, feel more caught in a state of flux. They still have a greater range of choices open. After a certain age threshold (pensioned status), willingness to get involved in reconstruction decreases.

Decision-makers in a situation of crisis should draw from this the lesson that it is important to deal with restoring employment early, because this produces a consolidating effect on a society under stress. We shall have to show later how such measures can also have a negative effect.

Through its positive or negative socialization functions, the sort of emergency housing used can exert an influence on victims of catastrophes. If the accommodation is too spartan, it increases emotional pressure on the occupants (because of such things as unsoundproofed walls and, in general, social controls impinging on privacy); but if they provide too high a living standard, perhaps superior to the dwellings that had been destroyed, they tempt people to become content with the condition they have reached and give up altogether the idea of permanent reconstruction.

Decision-makers in a situation of crisis ought not, under any circumstances, to set up emergency housing anywhere that varies markedly in quality, for this arouses envy or feelings of superiority in the occupants and heightens social tensions.

Notes

1 Burton, I., R. W. Kates and G. F. White 1978. *The environment as hazard*. New York: Oxford University Press.
2 Rijkswaterstaat (a Dutch authority) estimated costs of restoration at 1.1 billion hfl (= $500 million), and the total costs of the Delta plan at *c*. 6 billion hfl ($3 billion). See: Institute for Social Research in the Netherlands 1955. *Studies in Holland: flood disaster 1953*, 3 vols. Amsterdam.
3 For reconstruction in Friuli, $4.2 billion were alloted over five fiscal years by the law of August 8, 1977.
4 Di Sopra, L. 1977. *Stima dei danni*. Documento sulla ricostruzione, April 1.
5 White, G. F. (ed.) 1974. *Natural hazards – local, national, global*. New York: Oxford University Press.
6 Mileti, D. S. 1980. Human adjustment to the risk of environmental extremes. *Sociol. Social Res*. **64**(3), 327–47.
7 Saarinen, T. F. 1966. *Perception of drought hazard on the Great Plains*. Research Paper no. 106, University of Chicago.
8 Anon. 1970. Organizational and group behavior in disasters. *Am. Behav. Sci*. **13**(3).
   Lazarus, R. 1966. *Psychological stress and the coping process*. New York.
9 Ley, D. 1974. *The black inner city as frontier outpost. Images and behavior of a Philadelphia neighborhood*, 96. Washington, DC: AAG Monograph no. 7.
10 Nichols, T. C. 1974. Global summary of human response to natural hazards: earthquakes. In White, G. F. op. cit. (see note 5), p. 277.
11 Unesco 1976. *Intergovernmental conference on the assessment and mitigation of earthquake risk. Final report*. Paris, February 10–19.
12 Wallace, R. E. 1970. Earthquake recurrence, San Andreas Fault. *Geol. Soc. Am. Bull*. **81**. This paper assumes a recurrence period of 102 years for an 8.0 R earthquake on the San Andreas Fault.
13 White, G. F. op. cit., p. 222.
14 Unesco 1979. International Symposium on Earthquake Prediction, Paris, April 2–6.
15 Bowden, M. J. 1970. Reconstruction following catastrophe: the *laissez faire* rebuilding of downtown San Francisco after the earthquake and fire of 1906. *Proc. AAG* **2**, 22–6.
16 Durante, F. 1976. *Terremoti in Friuli*. Udine. This mentions the earthquakes of October 15, 1812 (Trevigniano M VIII), 1839 and 1841 (Carnia and Arta M VI), June 29, 1873 (Belluno M IX), June 24, 1889 (Tolmezzo M VIII), March 27, 1928 (Tolmezzo M IX), October 18, 1936 (Cansiglio M IX), April 26, 1959 (Tolmezzo M VII) and December 18, 1971 (Claut M VIII).
17 Borcherdt, R. D. (ed.) 1975. *Studies for seismic zonation of the San Francisco Bay region*. USGS Prof. Paper 941-A, p. III. Washington, DC.
18 National Academy of Science 1970. *The great Alaska earthquake of 1964*. Washington, DC.
19 Los Angeles County Earthquake Commission 1971. *San Fernando, California, earthquake, February 9, 1971*. Report, November.
   Murphy, L. M. (ed) 1973. *San Fernando, California, earthquake of February 9, 1971*, 3 vols. Washington, DC.
20 Nichols, D. R. and J. M. Buchanan-Banks 1974. *Seismic hazards and land-use planning*. USGS Circular 690, p. 1. Washington, DC.

21 Wallace, R. E. 1974. *Goals, strategy and tasks of the earthquake hazard reduction program*. USGS Circular 701, p. III. Washington, DC.

22 The Governor's Earthquake Council 1974. *Second report*. California Division of Mines and Geology, Special Publication no. 46, December.

23 Panel on the Public Policy Implications of Earthquake Prediction 1975. *Earthquake prediction and public policy*. Washington, DC.

24 White, G. F. and J. E. Haas 1975. *Assessment of research on natural hazards*. Cambridge, Mass.: MIT Press.

25 Adamic, M. O. 1979. *Posledice potresov leta 1976 v. SR Slovenij*. Ljubljana. This describes the consequences of the May 6 earthquake on the Slovenian side of the border.

26 Cf. the high rate of refusals in Jackson and Mukerjee's survey in San Francisco:
Jackson, E. L. and T. Mukerjee 1974. Human adjustment to the earthquake hazard of San Francisco, California. In *Natural hazards – Local, National, Global*, White, G. F. (ed.), 163, New York: Oxford University Press.

27 The undoubtedly Spartan quality of a questionnaire run off on a duplicating machine was assessed against the psychologically positive effect of a printed form. The Bavarian State Office for Data Processing generously took over the whole printing job. It had to be determined above all, however, that the data could be processed in due time. The fact that the Leibniz Computing Center (the data processing center for both universities of Munich and the Bavarian Academy of Science) was in the process of moving during the critical phase of the work unfortunately prevented us from having the help of the normal data processing personnel, for the problems being dealt with at the Department of Geography. Through the kindness of the office of the Prime Minister of Bavaria, it was possible to secure the official aid of the Bavarian State Office for Data Processing in this case too.

28 Occupants of prefabs in selected communes.

Within 'comuni disastrati' only (%)		Within 'comuni gravemente danneggiati' as much as (%)	
Meduno	29.7	Dogna	65.0
Travesio	29.3	Villa Santina	31.6
Ragogna	25.4	Zuglio	27.8
Frisanco	20.4	Preone	20.5
Cassacco	18.3	Tolmezzo	20.5
San Daniele	17.9	Raveo	18.6
Faedis	17.8		

29 Rate of return.

Size of commune	Population as of Dec. 31 1975	In prefabs in May 77	%	Forms distributed	Forms collected	%
500–1000						
Resiutta	500	432	86.4	157	98	62.4
Tramonti d. St.	500	432	38.5	181	72	39.7*
Tramonti d. Sp.	719	263	36.6	65	39	60.0*
Montenars	720	570	79.2	166	91	54.8
Clauzetto	742	502	67.7	101	100	99.0
Frisanco	761	155	20.4	73	21	28.7
Amaro	813	450	55.4	131	69	52.6
Lusevera	1 000	871	87.1	300	224	74.6
8 communes	5 938	3 505	59.0	1 174	714	60.8
1001–2000						
Bordano	1 065	841	79.0	291	73	25.0
Castelnovo	1 078	634	58.8	214	143	66.8
(Arba)	(1 231)	(70)	(5.7)	(24)	(23)	(95.8)*
Taipana	1 234	695	56.3	336	158	47.0
Chiusaforte	1 261	697	55.3	219	74	33.8

Size of commune	Population as of Dec. 31 1975	In prefabs in May 77	%	Forms distributed	Forms collected	%
Cavazzo Carnico	1 280	850	66.4	229	96	41.9
(Vito d'Asio)	1 414	986	69.7	317	0	0.0*
Cavasso nuovo	1 422	896	63.0	223	153	68.6*
Fanna	1 509	722	47.8	219	71	32.4*
Resia	1 654	1 389	84.0	487	275	56.4
Pinzano	1 699	986	58.0	237	159	67.0*
Treppo Grande	1 741	699	34.4	136	104	76.4
Sequals	1 812	1 023	56.5	253	115	45.4*
Attimis	1 880	755	40.2	332	160	48.2
Travesio	1 896	556	29.3	171	116	67.8*
Meduno	1 900	564	29.7	145	144	99.3*
14 communes	24 076	11 063	45.9	3 833	1841	48.0
2001–3250						
Forgaria	2 095	1 521	72.6	482	235	48.7
(Magnano in Riv.)	2 161	1 717	79.5	516	(109	21.1)†
Colloredo di M.	2 190	800	36.5	254	139	54.7
Moggio Udinese	2 490	2 123	85.3	573	293	51.1
Osoppo	2 543	1 457	57.3	1 103	271	24.5
Cassacco	2 630	480	18.3	93	79	84.7
Venzone	2 652	2 320	87.5	796	314	39.4
Pontebba	2 854	1 016	35.6	306	181	59.1
Nimis	2 916	2 270	77.8	711	290	40.7
Artegna	2 948	1 600	54.3	631	274	43.4
Trasaghis	2 951	2 030	68.8	778	349	44.8
Ragogna	3 019	766	25.4	244	71	29.1
Faedis	3 206	570	17.8	162	116	71.6
12 communes	34 655	18 670	53.9	6 649	2612	39.2
5000–12000						
Majano	5 468	2 150	39.3	806	317	39.3
San Daniele d. Fr.	6 857	1 225	17.9	317	251	79.1
Buia	6 865	4 050	59.0	929	151	16.2
Tarcento	9 314	5 859	62.9	1 757	188	10.7
Gemona	11 651	8 415	72.2	1 056	494	46.7
5 communes	40 155	21 699	54.0	4 865	1401	28.7
39 communes	104 824	54 937	52.4	16 521	6568	39.7
Udine Provine	87 958	47 317	53.8	14 298	5435	38.0
Pordenone Provine	16 866	7 620	45.2	2 223	1133	50.9
(Vito d'Asio)	1 414	986	69.7	317	0	0.0*

* = Pordenone Provine; † = forms lost; ( ) not evaluated.

30 Differences between Udine and Pordenone Provinces and between mountains and hills.

Question content	Udine	Pordenone	Mountains	Hills
1 communes	28	11	14	25
number of questionnaries	5435	1133	2539	4029
2 average age (years)	53.41	57.15	53.97	54.12
3 proportion of interviewed females (%)	36.3	38.1	34.7	37.8
4 average number of household members	2.15	2.03	2.12	2.14

Question content	Udine	Pordenone	Mountains	Hills
5 loss of near relatives (%)	13.9	2.6	10.4	13.0
6 loss of work place (%)	12.6	5.0	12.4	10.6
7 house ownership (%)	77.0	79.0	81.3	74.8
8 non-agricultural business (%)	12.8	14.3	13.4	12.8
9 land property in commune (%)	65.5	71.7	74.2	61.7
10 wage earner in Friuli (%)	53.1	49.7	52.4	52.7
wage earner abroad (%)	3.8	5.3	3.8	4.3
unemployed (%)	28.2	27.4	29:9	26.8
11 other work place than before earthquake (%)	21.1	18.3	22.1	19.7
12 pensioner in family (%)	63.8	70.3	66.0	64.2
13 earthquake in May (%)	36.0	25.9	34.4	34.1
earthquake in September (%)	3.9	2.8	5.7	2.4
both earthquakes (%)	46.4	63.9	47.7	50.5
14 reconstruction May (%)	14.1	9.8	16.4	11.4
no reconstruction (%)	68.1	74.1	69.6	68.8
15 remain in prefab (%)	46.2	28.3	44.7	42.2
reconstruction (%)	37.6	56.7	39.1	·42.0
move away (%)	0.2	0.2	0.2	0.2
16 type of house: wood (%)	17.2	15.4	16.8	17.0
17 life in a town (%)	10.0	5.9	8.5	9.8

31 Cf. *Ricostruire* 1(2), with a report on 250 interviews about nine types of prefabs in seven communes.

3 The earthquake effects

3.1 Damages, deaths and dislocation of the regional economy

The earthquake *events* have been described in Chapter 1, p. 1. What were their *effects*?

Even though with nearly 1000 people dead and 2400 injured, the complete destruction of 32 000 homes and damage to those of 157 000 more people, leaving 89 000 homeless in the provinces of Udine and Pordenone (Table 3.1), the regional economy nevertheless proved quite resilient.

The earthquake occurred north of the development axis of Pordenone–Udine–Gorizia–Trieste (cf. map on p. 11), where the level of socio-economic development was highest. This brings up the question of whether the regional economy would concentrate even more strongly on this axis after the catastrophe, for example because destroyed firms when they rebuilt would move directly there and abandon the disaster zone altogether.

Friuli is an area where commuting is well developed, because property in house and land is highly valued and keeps people in their home communes even when industrial employment is lacking. We will try to find out to what degree this hold may have been loosened.

The severest effects undoubtedly occurred where losses of life were registered, and these can be shown as the ratio of the number of dead to the total district population. Figure 3.1 shows the narrow concentration of areas worst stricken.

The next severest level of effects could be detected by finding out whether or not living quarters or working places, serving as the main reason for a desire to stay, had been destroyed.

(a) The worst case (work and living places both destroyed) applied especially to decision-makers whose existence rested on the traditional

Table 3.1 Persons without shelter, December 1976.

	Number of communes affected		
Homeless as % of resident population	*Udine Province*	*Pordenone Province*	*Total*
80–100	10	6	16
60–80	13	1	14
40–60	9	2	11
20–40	7	2	9
0–20	36	27	63
total	75	38	113
total communes in province	137	51	188

Source: ITALSTAT, *Piano-quadro di rinascità del Friuli*, Roma 1976, cf. Valussi (1977), p. 117 – see note 13, Chapter 5.

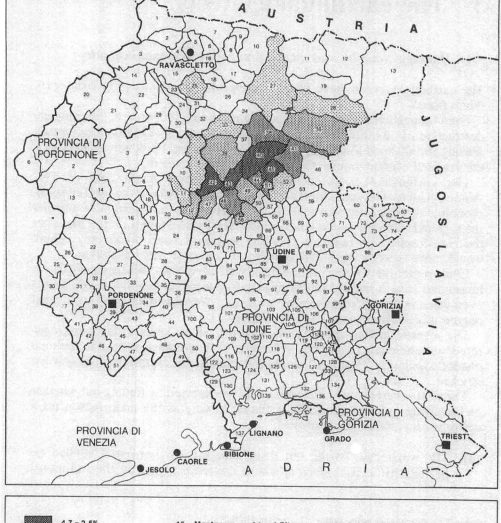

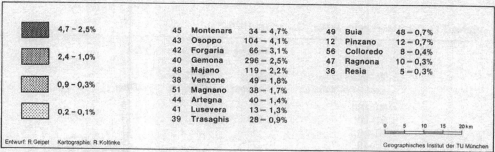

45	Montenars	34 = 4,7%	49	Buia	48 = 0,7%
43	Osoppo	104 = 4,1%	12	Pinzano	12 = 0,7%
42	Forgaria	66 = 3,1%	56	Colloredo	8 = 0,4%
40	Gemona	296 = 2,5%	47	Ragnona	10 = 0,3%
48	Majano	119 = 2,2%	36	Resia	5 = 0,3%
38	Venzone	49 = 1,8%			
51	Magnano	38 = 1,7%			
44	Artegna	40 = 1,4%			
41	Lusevera	13 = 1,3%			
39	Trasaghis	28 = 0,9%			

Figure 3.1 Earthquake deaths as a percentage of commune population.

economic pattern of 'home and work under one roof', such as small merchants, businessmen, handicraft workers, proprietors of inns and taverns, as well as doctors, lawyers and accountants with practices at home, and farmers. Within this group of 'independent professions' were included, however, some who would profit from the catastrophe, such as dealers in building materials, construction workers and repairmen, who could gain from reconstruction, as well as furniture and household supply businesses that could profit from the need for replacement items. Most others experienced a decrease in the number and purchasing power of their customers or clients, or the shift of the latter to unharmed competing firms.

(b) In the second case, with destruction of the home but survival of the place of work, two variations occurred with respect to mobility. The work place might be in the same commune as the destroyed home. Then the willingness to remain would depend on the quality of this work place and the likelihood of being able to restore the house within a reasonable time, in addition, as in every other case, to numerous imponderables – attachment to the home, loss of relatives and communication linkages. At least work gives a concrete basis for the restoration of the home. When a work place in *another* commune is preserved and the home destroyed, the commuter concerned would consider the cost of the journey to work to be higher than he did before, being able no longer to take for granted living in a given place where he had inherited property, family ties, and a network of communication. The decision-maker might decide to move to where he works. If this meant somewhere outside the earthquake zone, such as the Pordenone–Udine–Gorizia development axis, then the total effect of many such individual decisions would be an increased settlement density there, causing an upward movement in the cost of living and a scarcity of goods and services. Investment in reconstruction in the disaster area thus competed with investment in new starts for people moving into the development area.

(c) The third case (work place destroyed, home preserved) occurred in two variations, according to whether the destroyed work place was within the home community or outside. Temporary or lasting unemployment, change of occupation, a job beneath one's qualifications etc., were probably easier to take if one did not at the same time have to be wrenched out of an admittedly wrecked but still existing and therefore probably all the more prized home environment. By definition, we find destroyed work places only within the disaster area. Even when they were back in operation, they came into competition psychologically with work places that were really or supposedly more secure, for instance in the development axis. Possibly, it will turn out after they are re-established in the disaster area that in functional terms they have more modern equipment (which may mean lay-offs in the name of efficiency), or that they could be superior in human terms because the determination of the workforce, thinking first of the businesses and factories and only then of their own homes, might have improved the business climate.

(d) In the fourth case (homes destroyed), which would apply also to children and young people, the category of 'population no longer economically active' became important. The problem was defined for these people by the living situation. In order to replace a residence, far greater means were required than have ever been available to the victims in the entire course of their later lives. A large number of the older people who were housed at the end of the year 1976–7 in coastal-zone hotels converted into homes for the elderly, sometimes needing special care, had lost all their working relatives through death or emigration. It was hardly possible to imagine them going back to new wooden houses in the disaster area. Old people's homes or nursing homes would have to be built for them, which would involve putting an overall official team in charge of functioning arrangements to take care of individual cases from many different locations. But this would mean that many of the people could not return to their original communities.

A variation of case (d) can be conceived, in which many houses were already empty before the earthquake. Their owners had planned to come back to their home communities from abroad or other areas of Italy only after retiring. These now-destroyed provisions for their later years were characteristic mostly of more remote mountain villages.

The shift (see Fig. 2.4) of earthquake epicenters could be brought together with social science data, for the northward shift of earthquake ravages could be read off the maps of numbers of people homeless in the May and September eathquakes. (This northward shift was not sufficiently taken into account by the Italian government when the division was established according to degree of damage among 'comuni diastrati', 'comuni gravemente danneggiati' and 'comuni danneggiati'. The commune of Dogna, in particular, ought to have been included among the most severely damaged.)

The greater degree of damage in the mountain areas was also reflected in the fact that they showed an especially high proportion of older people as a result of the emigration processes described. Among the 187 districts in Udine and Pordenone Provinces, those with over 25 per cent of their inhabitants aged over 60 accounted for only 21.9 per cent of the total population (Figs. 3.2 & 3.3). But among the 41 'comuni disastrati' were 20 districts with a percentage of older people as high as 48.8 per cent.

Figure 2.10, showing the 'Percentage of commune population older than 60' (p. 60), which depicts the borders of the natural mountain and hill country units, as well as the area included in the 'comuni disastrati' category (identical with the areas where the questionnaire was administered, cf. Ch. 4), clearly reveals that, with few exceptions, it was only in the plains (pianura) that large contiguous blocs of districts occurred that fell under the average value for elderly inhabitants of both provinces (Udine, 20.0 per cent, Pordenone, 18.7 per cent) whereas the communities in the 'colline' and 'montana' natural land units mostly rose above it.

But this very situation of homelessness gave the first signs that the spirit of resistance of the people soon would be rekindled. An indicator of it was refusal to evacuate. Figure 3.4 shows the state of the evacuation as of Christmas 1976. The commune of Osoppo (no. 43 on the Commune Master List), for example, despite complete destruction, shows less than 40 per cent of its population participating in the evacuation. The industrial zone of Osoppo was the biggest center of employment in the catastrophe core area. The rapid reconstruction of enterprises destroyed here retained parts of the employable population and re-incorporated them into reconstruction with almost no interruption caused by halted production. The versatility of the Friulian workforce was in the main responsible for this, and the same employees now tended cement mixers or did masonry instead of attending blast furnaces and

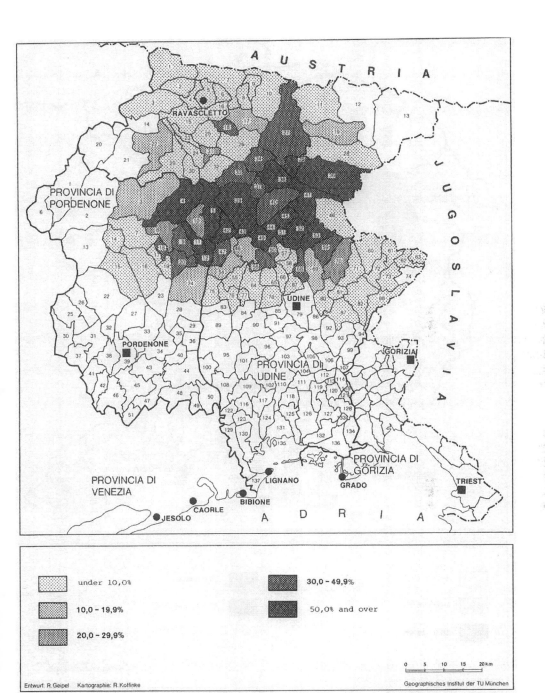

Figure 3.2 Persons left homeless on May 6, 1976 as a proportion of total population on December 31, 1975, by commune.

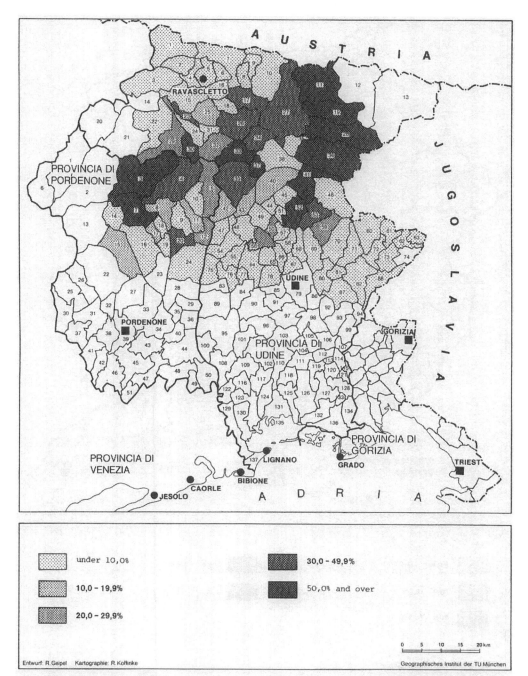

Figure 3.3 Persons left homeless on September 15, 1976 as a proportion of total population on December 31, 1975, by commune.

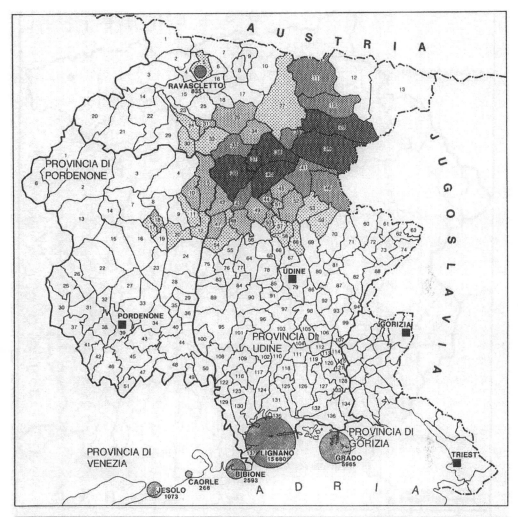

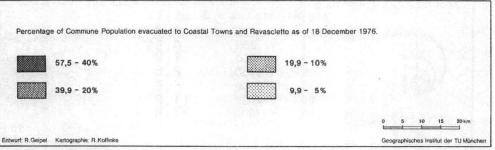

Percentage of Commune Population evacuated to Coastal Towns and Ravascletto as of 18 December 1976.

- 57,5 – 40%
- 39,9 – 20%
- 19,9 – 10%
- 9,9 – 5%

Entwurf: R.Geipel   Kartographie: R.Koffinke

Geographisches Institut der TU München

Figure 3.4 Population – evacuation. The situation as of December 18, 1976.

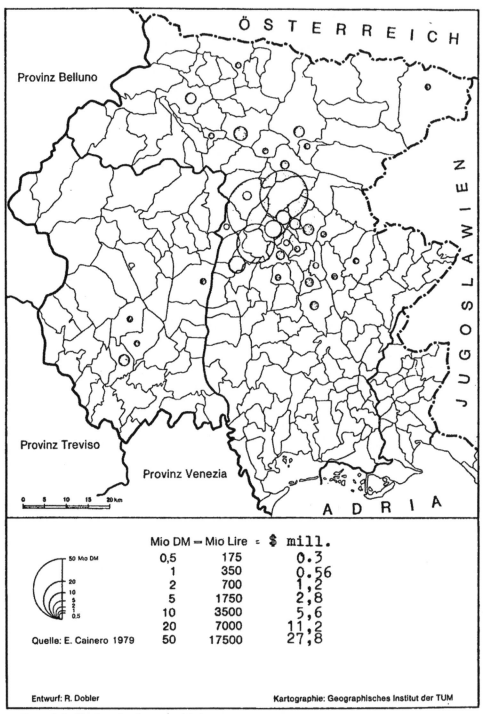

Figure 3.5 Earthquake damage to industry.

workbenches. This was supplemented by the government's continued payment of the normal wages of workers in firms that had stopped operating.

Figure 3.5 is an attempt to quantify earthquake damage to industry according to Italian sources. There were 326 damaged enterprises in Udine Province, with total damage of approximately $84.7 million, and in the Province of Pordenone 58 firms with $2.35 million damage. Some 20 000 people were employed in these enterprises, of whom 4000 lost their jobs. Only the four worst-hit firms can be cited here:

The Gemona Textile Factory	$20.6 million;
Fantoni Furniture Co., Osoppo	$14.2 million;
Pittini Steel Works, Osoppo	$14.2 million;
Snaidero Furniture Co., Majano	$10.0 million.

Apart from the textile factory in Gemona, which only began production again in 1978 after complete rebuilding, the interruption of production did not last

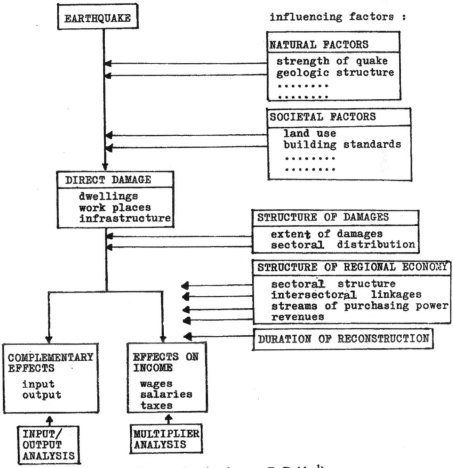

Figure 3.6 Economic losses from earthquakes (source: R. Dobler[1]).

Table 3.2 Sectoral apportionment of damage in industry in the Province of Udine.

	Mountain area Million lire	%	Central hill area Million lire	%	Udine/Cividale Million lire	%	Total Million lire	%
mechanical	58	2.2	4 476	7.6	101	7.6	4 635	7.4
metallurgy	26	1.0	13 776	23.5	357	26.7	14 159	22.7
wood and paper	2421	90.8	2 780	4.7	56	4.1	5 257	8.4
furniture	100	3.7	13 327	22.8	150	11.2	13 577	21.7
textiles and clothing	24	0.9	13 718	23.5	24	1.8	13 766	22.0
chemical	–		1 622	2.8	270	20.2	1 892	3.0
foods	8	2.9	2 724	4.7	42	3.1	2 773	4.4
construction	30	11.3	5 900	10.1	328	24.5	6 258	10.0
other	–		166	0.3	11	0.8	177	0.3
total	2666 (4.3%)	(100)	58 489 (93.6%)	(100)	1339 (2.1%)	(100)	62 494 (100%)	(100)

Source: Camera di Commercio di Udine (1977), with added information by R. Dobler (1980).

long. Even using Di Sopra's intentionally pessimistic estimates, it was clear that over three-quarters of the firms affected were fully operative again just three months after the quake.

Damage consisted not only of the destruction of buildings, machinery, transport installations, stockpiles, and so on, but also of the drop in production during the period the enterprise was inoperative. The number of lost work days in relation to the working year of various branches allowed R. Dobler,[1] of our department's research team, to quantify losses due to the cessation of production (Fig. 3.6). They were highest in the central hill country, with about 20 per cent of the yearly work level lost; metallurgical and mechanical industries suffered most, with a 27 per cent loss of annual production (Table 3.2).

Agriculture suffered the destruction of, or damage to, 28 000 individual enterprises, and 2250 head of cattle were killed, while 9500 head had to be transferred temporarily to farms in the plains. Among the many smallholders, especially, the loss of animals and facilities may have strengthened the trend toward giving up farming altogether that was already evident before the earthquake.

Damage to residences was estimated at around $300 million, and that to productive facilities at about $250 million. Up to May 1978, a total of $65 million had been allocated by the regional government to agriculture, which is another sign that initial damage estimates had been too high, since these $65 million included development funds for various projects and not just repair of damage to settlements.

3.2 Longer-lasting impact on the regional economy

Even when we know that around 250 000 people die every year from natural catastrophes and $25 billion in financial losses are incurred, not to mention $15 billion in relief and countermeasures,[2] these statistics still do not provide a means of quantifying the real consequences.

The first attempts to analyze the complex regional economic consequences of catastrophes were undertaken by Dacy and Kunreuther[3] in 1969 and Cochrane in 1974.[4] They were intended to predict the possible damage that an earthquake like the one of 1906 in San Francisco might produce today. An amount of $7 billion, with $6 billion in indirect damage effects, was reported. But even 10 years ago, Dacy and Kunreuther recognized that losses from catastrophes are often overestimated.

In the immediate aftermath of severe disasters, there is a general feeling within the stricken community that destruction is greater than it actually is. Radio, television and newspapers focusing on human-interest reporting convey a similar impression to the outside world. At the same time, state and local officials view the situation in its darkest light, and their sight is usually faulty. Destruction is hardly ever total, and sometimes it turns out to be small in comparison with early guesses. Corps of Engineers officials responsible for damage estimates have observed that actual losses usually run about one-third of the figures given in the earliest statements.[5]

The damage from the Alaska earthquake was thus at first set at $600 million, for example. A revision of this estimate by the same authorities nine months later only reached the sum of $311 million.

Damage estimates made on the spot and before the shock has worn off grossly exaggerate the extent of losses caused by natural disasters. Yet these early guesses frequently make headlines and are the ones remembered longest by the public. Overstatements of damage have at least two effects: (1) They make it appear that recovery was much faster than it actually was, and (2) they make it easier politically for the Federal government to dispense aid more generously than it otherwise might.[6]

An overestimation of the long-term socio-economic effects of the catastrophe as a rule goes along with the overestimate of damages. Despite the variety of cumulative effects, which can lead to a great increase in losses, the actual influence of natural catastrophes on regional development is often less than might be feared. Recent research in the United States even suggests that natural catastrophes exert no long-term effect at all on socio-economic development.[7] Such results may be true within the circumstances of the social framework of the catastrophes investigated, but in looking over the total existing material about development within disaster areas one is led to the suspicion that, with certain preconditions, positive consequences may even predominate. R. Dobler made a special point of this in his dissertation on Friuli.
As early as 1963, Fritz[8] pointed out this possibility for the first time.

The remotivation of the actors within the system and the consequent total concentration of societal energy on the goals of survival and recovery usually result in the rapid reconstruction of the society and, beyond that, often produce a kind of 'amplified rebound' effect, in which the society is carried beyond its pre-existing levels of integration, productivity, and capacity for growth.[9] (For example, the German postwar 'miracle' of reconstruction).

93

This indication of an accelerated upswing over the pre-catastrophe level was hardly picked up in the early years of hazard research. But it has gradually become clear that catastrophes can have eufunctional as well as dysfunctional consequences[10] for the societal system they affect. It remained uncertain just which influences played a role in bringing about these favorable or unfavorable consequences, however. One of the major objectives of Dobler's work within our Friuli team was, therefore, to clear up some of these complex relationships.

The master theme of the rest of this chapter will therefore be the question, to what degree may we regard catastrophes not as hindrances to development, but, on the contrary, as an impetus to development in the area they affect?

The ensuing empirical examples show that this at first astonishing hypothesis is really altogether plausible. It must be emphasized, however, that the hypothesis refers to processes at an aggregate, supra-individual level on the plane perhaps of regional economy, but not to individual hard-hit victims of catastrophe who would view such a pronouncement as pure cynicism.

Dacy and Kunreuther cited the explosion disaster in Halifax and the Alaska earthquake, among others, as evidence for this hypothesis. Herweijer[11] was able to show the same thing in the case of the great flood catastrophe in the Netherlands. On that day, February 1, 1953, 1787 people died in the flood, 141 000 ha of arable farmland were ruined, and total losses were estimated at $3 billion. The Dutch Government assumed 90 per cent of the costs, completely rearranged the agrarian structure and, by means of multifunctional protective works in the Delta-Plan, laid the groundwork for the Province of Zeeland to become one of the most modern and productive in the country.

In the first weeks following the disaster in Friuli, an Italian team led by the architect and planner Luciano Di Sopra attempted an estimate of damage,

Table 3.3 Losses from earthquakes and floods (* = Europe).

	Year	Casualties	Damage ($ million)
Earthquake catastrophes			
San Francisco	1906	750	524
*Messina	1908	83 000	?
Yokohama	1923	142 800	2800
Agadir	1960	13 100	120
*Skopje	1963	1 070	300
Alaska	1964	131	538
*Val Belice, Sicily	1968	281	320
Managua, Nicaragua	1972	5 000	800
*Lice, Turkey	1975	2 400	17
Guatemala	1976	22 800	1100
Tang-shan, China	1976	665 000?	?
*Bucarest, Romania	1977	1 581	800
Flood catastrophes			
*Netherlands	1953	1 787	3000
*north-west Germany	1962	347	600
Karachi, Pakistan	1965	10 000	?
Bangladesh	1970	300 000	63
USA (hurricane)	1972	122	3100

published in April 1977.[12] It told of losses at a scale of 3420 billion lire, or $4.75 billion as of 1976. Of this sum, the production sector accounted for $1.74 billion and settlements for $2.8 billion. Table 3.3 compares these $4.7 billion with loss estimates for other great natural disasters, as they were listed by one of the largest insurance companies in the world, the Munich Re, in 1978. This comparison makes it clear that local Friulian politicians, claiming a damage of $4.750 billion, exaggerated the magnitude of the disaster for the benefit of the government in Rome, probably in order to compel the state to help in the catastrophe area with a comprehensive transfer of funds.

Di Sopra's estimate of losses is broken down as shown in Table 3.4.

This exaggeration of figures by those with concern for the Friulians was understandable because, after the earthquake of 1968 at Val Belice in Sicily, relief supplies from Rome soon went astray and were steered into obscure channels, leaving the catastrophe area hardly changed at all after 12 years. The Friulians had this example clearly before them, and it was strengthened by contacts with Sicilian earthquake victims (cf. Fig. 5.18). One of Dobler's models shows what influential factors may modify the consequences of catastrophe (Fig. 3.7).

Meanwhile, the Italian regional authorities conducted a survey of 271 damaged enterprises in order to check on these figures. They take a more optimis-

Table 3.4   **Di Sopra's estimate of losses.** These losses from the earthquake of May 6, 1976 supposedly were increased 20–30% by the one on September 15. Thus the figure of nearly $6 billion was arrived at.

	Losses ($ million) of 1976
Losses in productive sector	1745
direct losses	466
agriculture	281
industry	133
handicraft	25
trade and services	27
indirect losses	1279
decline in production	846
effects on incomes	156
inter-regional linkages	278
Losses in settlement system	2804
damage to dwellings	1826
destroyed	553
partially destroyed	960
renovation measures needed	238
outbuildings damaged	57
interior furnishings	22
damage to public installations	978
technical infrastructure of communes	66
social infrastructure of communes	389
regional infrastructure of communes	93
cultural properties	432
Hydrogeologic damage	201
total	4750

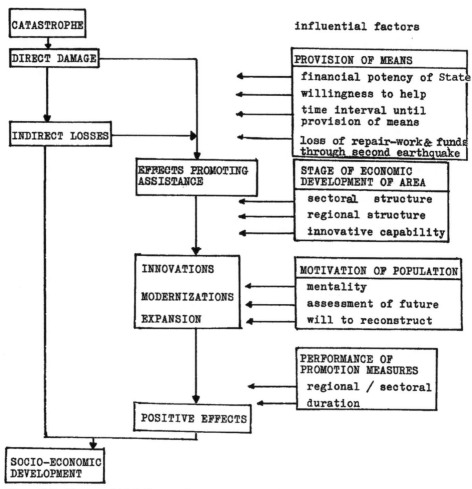

```
CATASTROPHE                                    influential factors

DIRECT DAMAGE ─────────────────┐   ┌─────────────────────────────────┐
       │                       │   │ PROVISION OF MEANS              │
       │                       │   │ ◄── financial potency of State  │
       │                       │   │ ◄── willingness to help         │
       ▼                       │   │ ◄── time interval until         │
INDIRECT LOSSES ───────────────┼──►│     provision of means          │
       │                       │   │ ◄── loss of repair-work & funds │
       │                       ▼   │     through second earthquake   │
       │           ┌──────────────────┐ ┌───────────────────────────┐
       │           │ EFFECTS PROMOTING│ │ STAGE OF ECONOMIC         │
       │           │ ASSISTANCE       │ │ DEVELOPMENT OF AREA       │
       │           └──────────────────┘ │ ◄── sectoral   structure  │
       │                   │            │ ◄── regional structure    │
       │                   │            │ ◄── innovative capability │
       │                   ▼            └───────────────────────────┘
       │           ┌──────────────────┐ ┌───────────────────────────┐
       │           │ INNOVATIONS      │ │ MOTIVATION OF POPULATION  │
       │           │ MODERNIZATIONS   │ │ ◄── mentality             │
       │           │ EXPANSION        │ │ ◄── assessment of future  │
       │           └──────────────────┘ │ ◄── will to reconstruct   │
       │                   │            └───────────────────────────┘
       │                   │            ┌───────────────────────────┐
       │                   │            │ PERFORMANCE OF            │
       │                   │            │ PROMOTION MEASURES        │
       │                   ▼            │ ◄── regional / sectoral   │
       │           ┌──────────────────┐ │ ◄── duration              │
       │           │ POSITIVE EFFECTS │ └───────────────────────────┘
       │           └──────────────────┘
       │                   │
       ▼                   ▼
┌──────────────────┐
│ SOCIO-ECONOMIC   │
│ DEVELOPMENT      │
└──────────────────┘
```

Figure 3.7 Factors which influence the consequences of a catastrophe.

tic view today of losses incurred due to interruption of production, for three months after the earthquake more than three-quarters of the affected enterprises were already operating again (Table 3.5).

Figure 1.8 shows Di Sopra's classification of zones A, B, C and D according to degree of damage. Losses were distributed as shown in Table 3.6, according to his account.

The reported sum of approximately $159 million in losses can be reduced to about $87 million. The overestimate of loss by reduction of output was vastly greater, however. The latter can best be expressed for industry in terms of the number of work days lost (Table 3.7).

Dobler's table shows that even in the central hill country, which was the zone of severest damage, less than 20 per cent of annual working time was lost as a consequence of the earthquakes. In order to evaluate these figures correctly, Friulian workers' industriousness (which is rather uncommon in the Italian context) as well as their enormous work discipline and their great

Table 3.5 Interruption of production, according to damage area.

	No interruption	0–1 month (%)	1–3 months (%)	3–6 months (%)	Still closed, May 1977	n
destroyed area	6.7	55.8	13.3	11.7	12.5	120 (100%)
severely damaged area	13.4	61.2	1.5	19.4	4.5	67 (100%)
damaged area	31.0	64.3	2.4	2.4	–	84 (100%)
total	15.9	59.8	7.0	10.7	6.6	271 (100%)

Source: Regione Friuli–Venezia–Giulia, Ass. Industria 1977 A. Table 3.

Table 3.6 Earthquake damage to industry, zoned by extent of losses.

Degree of damage (by zone)	Employees in plants affected Absent	% of total labor force	Direct losses In $ million	In % of capital
A	6 200	99	114	64
B	6 539	60	36	11
C	5 663	11	7.6	0.5
D	450	0.2	1.7	0.2
total	18 852	20	159.3	5

Source: Di Sopra (1977), p. 14.[12]

Table 3.7 Reduction in working days, by type of industry and socio-economic zone.

	Mountain zone Lost workdays Abs.	In %*	Hill zone Abs.	In %	Zone Udine/ Cividale Abs.	In %	Province of Udine Abs.	In %
mechanical/ metallurgical	1 000	0.5	104 700	27.2	23 400	1.2	129 100	4.5
wood products, paper, furniture	24 400	6.1	63 300	18.4	11 300	0.8	99 100	4.2
textiles/clothing	200	7.0	58 300	14.0	1 600	0.3	60 100	4.2
chemical	–	–	3 800	7.3	2 000	0.4	5 800	0.9
food products	100	0.8	6 500	9.1	300	0.1	6 900	1.7
construction	300	0.1	35 200	19.7	14 200	1.7	49 700	3.1
total	26 000	2.9	271 800	18.9	52 900	1.0	350 600	3.8

*Proportion of total yearly working days for each type of industry.
Source: Camera di Commercio di Udine (1977), completed by R. Dobler.

loyalty to their companies, have to be taken into account. Being skilled do-it-yourself craftsmen, they switched over from their destroyed regular work places and themselves repaired the damage to industrial plants. They knew they could only remain in their home area if they restored industrial work places as rapidly as possible.

While their dwellings lay still in ruins (which is mostly the case even now, four years after the quake, in 1980), the rebuilding of industrial plants was accomplished in an astonishingly short time. Chapter 4 details this process, on the basis of first-hand surveys.

This hesitation about civic rebuilding is readily understandable. Decision-making power with regard to communal facilities and to industrial plants was in the hands either of the state, of mayors, or of individual entrepreneurs.

Friuli has produced many entrepreneurs who began on a small scale, as self-made men. Their origins in the local handicraft system keep them close to the ordinary working man. Some of them are still only first-generation entrepreneurs, not afraid to work with their hands, unlike some spoiled heirs of industrialists. A Friulian is inclined to avoid incurring debts or taking heavy risks. Nevertheless, a whole series of entrepreneurial figures have taken their chances on this step from the family handicraft enterprise up to small- or middle-scale industry. When the earthquake destroyed the plant and disrupted deliveries to customers, since the catastrophe was widely publicized, national and international credit institutions could be counted on to be sympathetic and a new start was chanced. State subsidies and low-cost loans promoted additional investment and increased the number of jobs.

The ordinary people, on the other hand, were pushed into the background while reconstruction of dwelling space was going on, so that, for them, the after-effects of the catastrophe have been getting worse instead. The previous property pattern and power structure came back all the stronger in the reconstruction phase, and a critic who claimed that 'an earthquake is a social class quake' may be wrong in the case of Friuli. This can be seen in the way better contacts and the possession of superior information permit influential entrepreneurs of the communes to commence reconstruction immediately, whereas the little man trying to rebuild his home on his own is put off with the claim that at the moment building capacity is completely tied up.

While he is getting the runaround this way, he becomes caught in a double bind, for the inflation rate of approximately 20 per cent a year eats up the reconstruction indemnity he has been counting on. At the same time, wages and the cost of materials are increasing steadily so that, for the victim of inflation, the fondest dream of a Friulian, living in his own home again, is shoved off farther and farther into the future.

This delay is of course nobody's fault in particular. It is a side-effect of the fact that when settlements are being rebuilt – unlike industries – a large number of decision-makers with all sorts of capital backing, varied plans for the future, and different ideas about construction, have to be brought together before a definite building plan can be put into effect. Even where the laws of the land are designed to insure equity and justice toward everyone who has suffered losses, socio-cultural and politico-economic variations in the ability to take advantage of these laws provide an opening for injustices. The people less informed and less capable of making important decisions, therefore, need help in situations of catastrophe in the form of advocacy planning.

3.3 Regional differentiation of the earthquake effects

Since our investigations spanned four years (1976–9) and employed a succession of different methods in order to determine the areal differentiation of damage, we can offer some insight into the temporal course and after-effects of a catastrophe. On p. 34 (Fig. 1.17) we showed how severity of an earthquake, as registered through indicators such as number of deaths, persons homeless, number of evacuees, and occupancy of barracks, is areally differentiated.

Our 1977 questionnaire covering 6500 households with more than 20 000 residents showed, on the basis of personal data, that the social predisposition of a commune to overcome earthquake consequences was destined to be important in the future, as marked by indicators such as an elderly population, a large proportion of females, retirement status, and numbers working abroad.

Three years later, R. Stagl[13] extended the study, analyzing the reconstruction process and the influence of size and the spatial location of a commune on the results of reconstruction and related difficulties, by means of 95 interviews with the planners in the communes totally destroyed.

The results of these three surveys can be incorporated into a three-dimensional co-ordinate scheme (Fig. 3.8).

It was revealed that the 'difficulties' indicated by many kinds of evidence do not correspond simply with the extent of destruction. They are, in fact, much more severe in small, remote communes than the extent of destruction in them would suggest (Fig. 3.10). This 'difficulty of problems' was arrived at by combining the factors of spatial position ($\times$ 1), degree of destruction ($\times$ 1.5) and size of commune ($\times$ 2), giving the results in Table 3.8.

Transforming the table into a map, we see that in the aftermath of the catastrophe the problems were displaced from the central disaster zone, represented by the map of 'Near relatives lost' of 1976 (Fig. 3.11), to outlying and structurally weak peripheral districts, represented by the map of 'Areal distribution of problems' of 1980 (Fig. 3.12).

The influences mentioned on p. 93 show up clearly in the process of successful, or unsuccessful, reconstruction.

In Type-E communes (Table 3.9), the advantages of agglomeration increased. They became growth centers, and the industrial plants were mod-

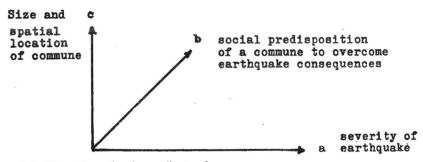

Figure 3.8 Three-dimensional co-ordinate scheme.

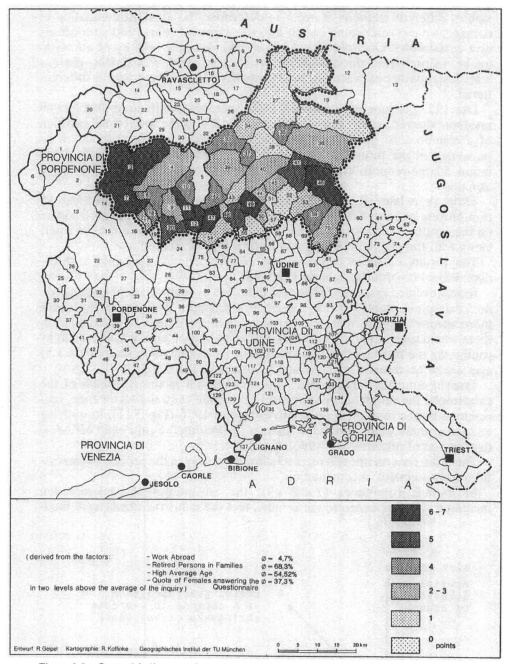

Figure 3.9 General indicators of structural weakness.

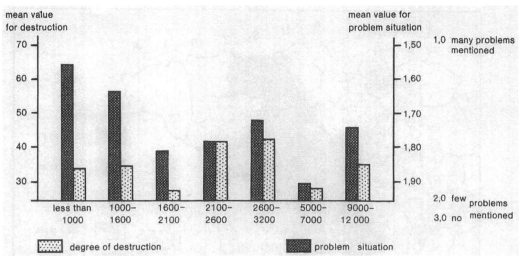

Figure 3.10 Comparison of extent of destruction and problems experienced.

Table 3.8 Categories of difficulty of problems.

Category I

(1) Resiutta	62	
(2) Montenars	61	
(3) Amaro	61	
(4) Clauzetto	61	
(5) Bordano	60	

(6) Lusevera	60	
(7) Vito d'Asio	60	
(8) Frisanco	60	
(9) Tramonti di Sopra	60	
(10) Tramonti di Sotto	60	

Category II

(1) Cavazzo Carnico	59
(2) Chiusaforte	59
(3) Taipana	59
(4) Trasaghis	59
(5) Venzone	59
(6) Castelnovo	58
(7) Cavasso Nuovo	58
(8) Fanna	58

(9) Forgaria	58
(10) Moggio Udinese	58
(11) Artegna	58
(12) Gemona	58
(13) Tarcento	58
(14) Tolmezzo	58
(15) Pontebba	58

Category III

(1) Resia	57
(2) Magnano	57
(3) Cassacco	57
(4) Faedis	57

(5) Nimis	57
(6) Ragogna	57
(7) Spilimbergo	57

Category IV

(1) Colloredo	56
(2) Osoppo	56
(3) Villa Santina	56
(4) Meduno	55
(5) Attimis	55

(6) Pinzano	55
(7) Sequals	55
(8) Travesio	55
(9) Treppo Grande	55

Category V

(1) Buia	53
(2) Majano	53

(3) San Daniele	53
(4) Tricesimo	53

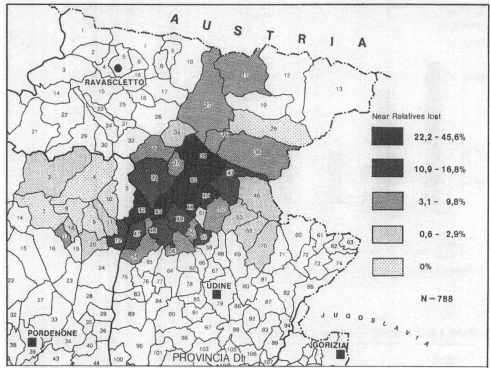

Figure 3.11   Near relatives lost, 1976.

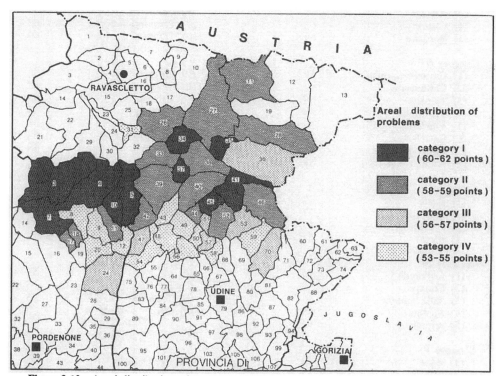

Figure 3.12   Areal distribution of problems, 1980.

Table 3.9 Types of reconstruction. After Stagl (1980).[13]

Indicators	Type A	Type B	Type C	Type D	Type E
population movement 1951–67	– – very strong decrease (55%)	– strong decrease (45%)	0 medium decrease (35%)	+/0 medium decrease (38%)	+ + stagnation (5%)
percentage aged more than 60	– – very high (35%)	– high (25%)	+ little (20%)	+ little (20%)	+ + very little (15–20%)
persons employed in primary activities	– – very high (25–30%)	0 medium (10%)	+ + very little (5%)	– high (20%)	+ little (5–10%)
employees in secondary activities	– – very little (45%)	+ high (60%)	+ high (60%)	– little (50%)	0 medium (55%)
initial situation	very bad – –	bad –/0	good +	medium 0	very good +/+ +
problem factor	– – very high (60 pts)	– high (59 pts)	0 medium (57–8 pts)	+ small (55 pts)	+ + very small (53 pts)
percentage in prefabs March 1978	– – very high (70%)	–/0 rather high (60%)	– high (65%)	+ low (35%)	+ + very low (25%)
degree of perplexity	very high – –	high –	rather high –/0	low +	very low + +
citizen satisfaction 1 = high; 5 = low	very low (3.56)	low (3.25)	very low (3.83)	very high (1.50)	high (1.80)
earthquake danger	– – very high	– – very high	0 medium	+ low	+ + very low
probability of another quake (planners' opinion)	– – very high (70 years)	– – very high (70 years)	– high (80 years)	+ + very low (100 years)	+ low (90 years)
comparative index of progress in reconstruction 1 = bad; 5 = good	– – very bad (1.67)	– – very bad (1.38)	+/0 medium (3.75)	+ good (3.90)	+ + very good (4.00)
prospects	very bad – –	very bad – –	bad –	good +/+ +	good +/+ +

ernized. In the communes of Type D, mostly bordering on them, incoming reconstruction supplies improved, or (in Type C) stabilized the situation. The losers from the catastrophe are communes of Type B and, particularly, A. Both the edges of the earthquake district, i.e. the mountain communes of Pordenone Province and the area bordering on neighboring Yugoslavia, still bear the worst scars of the catastrophe four years later. Their patterns can be appreciated by examination of Figures 3.13 and 3.14.

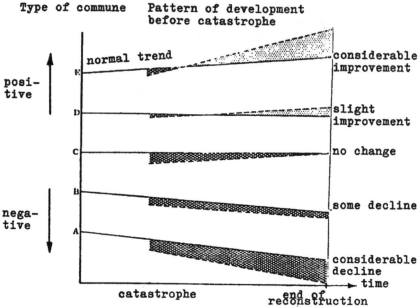

Type of commune   Pattern of development
                   before catastrophe

Figure 3.13 Trends of development.

3.4 Migratory tendencies and the readiness to return home

Migratory tendencies. Examining the trends of population in Friuli over the span of *one* generation, one sees a certain stabilization following the high rate of emigration in the 1950s, which is due to the high birth rate, but the latter is gradually decreasing (Table 3.10).

Return migration from abroad and other parts of Italy is noticeably on the increase – a sign that industrialization is also catching up in this outlying area of northeastern Italy. But this growing positive balance shown by Friuli–Venezia Giulia overall does not affect the whole region equally. A reapportionment is taking place in the area, on the contrary, to the advantage of the progressive, more heavily industrializing plains, and to a lesser degree the hill country, but very much to the disadvantage of the mountains.

Looking at the process over a period of *two* generations (1921–71), a dramatic development becomes apparent (Table 3.11).

So, while the population as a whole is stagnating in both Friulian provinces of Pordenone and Udine, it is being displaced out of the mountains into the plains. The axis of the process of emigration and relocation is right in the southern margin of the catastrophe zone. One might therefore suppose that the occurrence of the catastrophe would have strengthened this split, and that the more severely affected northern area would become more empty still. But, as the first population statistics from after the disaster reveal, this is not the case.

At the time of our May 1977 inquiry, only 0.2 per cent of the 6500 interviewees were disposed to leave Friuli for good, and 9.3 per cent would move to a large city. Even if statistically measurable, manifest mobility has not increased noticeably since then, it may nevertheless be assumed that latent

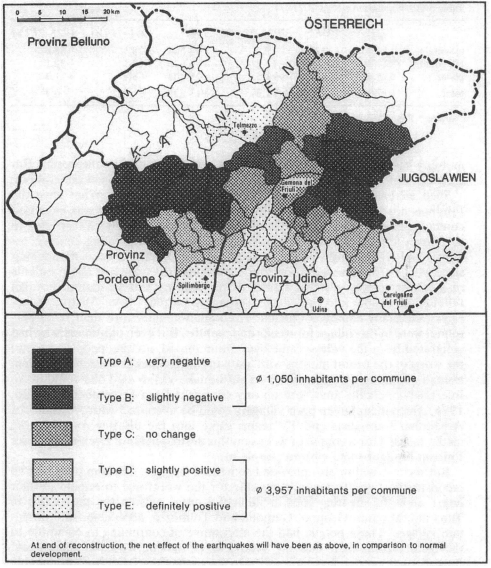

At end of reconstruction, the net effect of the earthquakes will have been as above, in comparison to normal development.

Figure 3.14 Trends of development shown areally.

Table 3.10 Population trends 1951–75.

	1951–61	1961–71	1971–75
balance of births	+27 856	+23 819	+2 206
migration balance	−49 679	−14 585	+28 815
total	−21 823	+9 234	+31 021

Table 3.11 Population trends 1921–71.

	1921	1951	1921–51 (%)	1971	1951–71 (%)	1921–71 (%)
mountains	151 856	132 678	−12.6	93 504	−29.5	−38.4
hill country	249 801	221 490	−11.3	194 620	−12.1	−22.1
plains	515 475	574 950	+11.5	625 104	+8.7	+21.3
total	917 132	929 118	+1.3	913 228	−1.7	−0.7

Source: Prost (1977), p. 91.

mobility has been slightly increased by the occurrence of the catastrophe. But it is *inter-regional* and no longer leads to temporal or permanent emigration.

Such increased, if still latent, mobility is not surprising. What have the Friulians not been through since the catastrophe! First there was increased contact with outsiders. The circle of experience which had been restricted to their own fellow villagers, companions at work in commuting centers, relatives and friends from nearby villages, has undergone a great broadening since May 1976: the evacuation to the coast into the luxurious environments of a leisure society set new standards for housing and sanitation. Foreign relief people came into the village, setting up barracks from Austria, Yugoslavia, Germany and even Canada. The Italian military with their relief personnel were in the village for weeks and months. But even old-timers who had emigrated from the village came back from abroad, inviting people to spend the worst of the winter months with a son in Switzerland or a sister in Belgium instead of among the ruins. With an earthquake-victim card one could make free telephone calls anywhere on any continent from September 15 to 30, 1976, which meant even poor villagers could be in contact with Australia, or Argentina. Journalists and TV teams came into the disaster zone and the media focused for a couple of weeks on this destroyed land. Even earthquake tourism brought some curious people in.

But reconstruction also provided some new options. The aim of industrial reconstruction was to make it possible for the workforce to remain in their home areas by creating jobs in industrial areas outside the mountains in Tarcento, Majano, Osoppo, Gemona and Tolmezzo, accessible from mountain villages. These people had the alternative of continuing to commute to the factory from their outlying valleys or of moving closer to their work. Reconstruction laws, however, failed to strengthen the latter highly desirable process, because they directed the indemnities to the commune of residence. We tried with our questionnaires to gain some insight into how the people were thinking about these matters.[14]

Since it is hard, by direct questioning, to gauge a potential tendency to move away at a future time, our team member, Ursula Wagner, inquired about how people would probably act when searching for a job. More than two-thirds of all 238 respondents, carefully grouped in six sets according to distance, would not move in any case, 20 per cent were prepared to move within Friuli, 2 per cent to other parts of Italy, and 6 per cent abroad. This mobility potential of 28 per cent seems larger than is suggested by the answers of 16 per cent who do not wish to proceed with reconstruction in their old communes (Table 3.12). Many old people, especially, are included in this 16 per cent. Since there were indemnities only for the former home in the old commune,

Table 3.12 Mobility potential.

	Reconstruction already begun (%)	Reconstruction planned (%)		Not applicable/ already in own home (%)	
		yes	no		
willing to move	20.8	51.5	16.0	11.7	$n = 231$

the decision to rebuild does not reveal whether or not the respondent wishes to remain permanently in the old mountain village. Probably the new house will serve as a weekend or vacation place for the later generations. This tradition was already established among the emigrants – property at home was not given up.

But even if people preferred at first to stay in the mountain village, the older people's decision was not binding on the children, particularly since, with the declining trend in agriculture, the latter would in the future be even more strongly compelled to go to work outside the village. Therefore, desired and actual daily journey to work was investigated (Table 3.13). Almost 90 per cent want a job within 45 minutes journey of their village. Nearly 30 per cent, though, must travel longer – up to 1½ hours. Their home communes also do not afford them the kind of facilities and services they consider necessary.

Among the 12 facilities our inquiry asked about, not even a school and kindergarten were found in every mountain commune, although the needs of 60 per cent of the respondents embraced at least nine items (Fig. 3.15).

Even though demand for such services will become more and more urgent for the next generation and will cause young parents especially to move out of mountain villages, the Friulian is nonetheless very much attached to the existing pattern of settlement, being no 'big city person', although two-thirds are still ready to live in places up to town size, including the (younger) employed people (Table 3.14).

Since the 211 respondents came from two groups of almost equal size representing central and peripheral portions of communes (Italian: *frazioni*), it is possible to detect the increase in dissatisfaction that goes with ever greater distance from the centers of activity: whereas only 19.3 per cent of people in the better-serviced 'central portions' would move to a bigger city or its suburbs, 26.5 per cent from the 'peripheral portions' would do so; 33.9 per cent of people in the central fractions but only 11.8 per cent in peripheral ones are content with the size of their own settlements. This will lead to a growing functional differentiation among various settlement sizes and a shift in population in the development of settlement structure in the longer term. Dobler's proposals are shown in Table 3.15.

Table 3.13 Desired and actual daily journey to work.

	Up to 30 min (%)	30–45 min (%)	45–60 min (%)	60–75 (min %)	75+ min (%)	(n = 119)
actual	58.2	12.7	8.9	6.3	13.9	reality
considered acceptable	70.6	18.5	8.4	0.8	1.7	desire

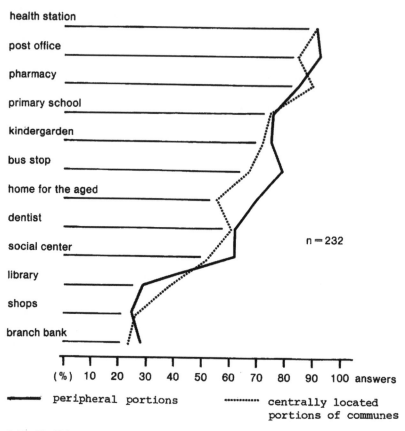

health station
post office
pharmacy
primary school
kindergarden
bus stop
home for the aged
dentist
social center
library
shops
branch bank

n = 232

(%) 10 20 30 40 50 60 70 80 90 100 answers

━━━ peripheral portions   ·········· centrally located
                                            portions of communes

Figure 3.15 Facilities needed in a commune.

Readiness to return home. In our study, we ran into the question on various occasions of whether the inclination of Friulians who had emigrated temporarily or permanently to come back home had been increased or decreased by the earthquake. Information on this point was needed for the planning of the building aspect of reconstruction, because space for potential returnees would have to be reserved and would at times be easier to secure with their

Table 3.14 Preferred place to live if move necessary.

	Bigger city (%)	Vicinity of bigger city (%)	Town (%)	Small town (%)	Central portion of commune (%)	Peripheral portion of commune (%)	n
employed	4.5	17.1	18.0	24.3	27.0	9.0	108
other respondents	5.7	17.0	9.4	16.0	34.9	17.0	103
central fractions	4.6	14.7	17.4	14.7	33.9	14.7	109
peripheral fractions	5.9	20.6	10.8	21.6	29.4	11.8	102
total	5.2	17.5	14.2	18.0	31.8	13.3	211

Table 3.15 Development of settlement structure – future functional differentiation.

Type of place	Services	Housing	Work
A = medium supply centre	increase in central place role with creation of mid-level installations	increase in residents by migration from C and D	creation of additional jobs in trade and services
B = first necessities centre	complete provision for basic daily supplies	increase in residents by migration from C and D	creation of additional jobs in handicraft and services
C = settlement core	improved supply situation through concentration of facilities in core	increase in residents by migration from D	maintenance of jobs in trade; new jobs in agriculture and transportation
D = other portions	supplied by C primarily and by mobile shopping services	decline of population migration to A, B and C	maintenance and increase of jobs in agriculture and transportation

financial help. The entrepreneurs also counted on their countrymen's willingness to come back, in view of the special shortage of skilled workers. What sort of factors influence and what can determine the return of workers employed abroad to their homeland?

Helene Voelkl,[15] one of our team members, sent 560 questionnaires to readers of the monthly *Fogolar* magazine in Germany and Switzerland, of which 141 (27.2 per cent) came back in usable condition.

The results of the study tested the following hypotheses.

(a)   The more integrated the emigrant is abroad, the less inclined is he to return. ('Integration' was measured by indicators such as circle of friends and acquaintances, frequency of meeting with same, knowledge of the language of the receiving country, economic and living arrangements abroad.)

(b)   The closer his contact with the homeland, the sooner he will return. (This relationship seemed to be gauged best by frequency and length of visits home, correspondence with home, and remittances of money or goods.)

(c)   The more information he has about home, the sooner he comes back. (What was known about economic development in Friuli and assessment of the earthquake risk seemed usable as indicators of how much the emigrant cared about having a realistic picture of his homeland.)

(d)   Return propensity is determined by socio-economic factors. (Indicators were age, sex, family status, household size, home or property ownership, place of origin, education, occupation and income.)

(e)   The fact that the original reasons for leaving Friuli no longer exist does not yet mean return. (What lay behind the emigration, stages that it included, last type of employment at home, last type of job held there.)

The following diagram (Fig. 3.16) shows factors influencing return home and the significance of relationships thus revealed. The fact that not one of

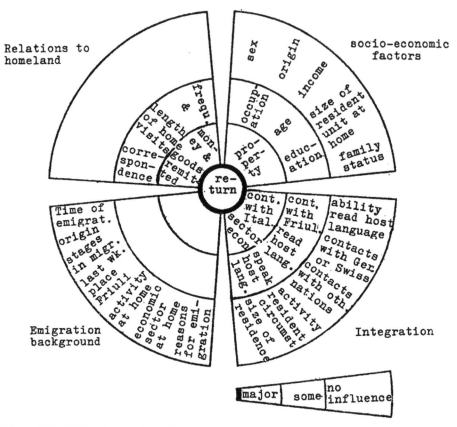

Figure 3.16 Factors influencing return home.

the 141 respondents wanted to stay abroad for the rest of his life is an indication of how attached the Friulians are to their homeland. Still, 78 per cent were unable to stipulate yet when they would return; 10 per cent wished to do so in 1980, 12 per cent between 1980 and 1986. The reasons for the planned return were various (Table 3.16).

If the respondents abroad became unemployed, 67.7 per cent would return home, and only 32.3 per cent would look for another job elsewhere. The inclination to return home is notably more pronounced among younger emigrants (Table 3.17). In contrast, the inclination to return soon decreases as the level of education increases. Emigrants who went farther in school, and know languages, take up more demanding work abroad, have undergone a longer

Table 3.16 Reasons for the planned returns.

	%
because of economic improvement in Friuli	39.0
in order to rebuild destroyed or damaged property	18.7
family reasons	17.9
total	75.6

Table 3.17 Inclination to return home.

Would take up an equivalent job at home	Age		
	Older than 50	40–50	Younger than 40
immediately	28.2	35.2	36.6
probably	10.0	42.0	48.0
no	41.7	41.7	16.7

apprenticeship, and therefore earn more and are not so ready to give up the positions they have attained (Table 3.18).

The particulars of the potential returnee can be described as follows, according to these and other statistical tests. He is under 40, has slight or medium education, has some destroyed or damaged property at home, and works either as a mechanic or in construction. In such a case, he has little chance to acquire what is imperative for integration into the host country – mastery of the language (Table 3.19).

Figure 3.16 shows that the actual background of emigration has no influence on propensity to return home. Social integration is also more important than economic: even with good living conditions and income in the host country, strong contact with Friulians and Italians indicates the feeling of being a stranger abroad. Encouraged in the 'duty' to go home by visits there, correspondence and remittances of money and goods there, and most of all by property destroyed in the earthquake, the people who first choose to go are mostly those who are most urgently needed at home – construction workers and mechanics (Table 3.20).

The most highly trained and best integrated abroad still hesitate, however. For them, Friuli must first lose its position as an 'outlying workbench' of the North Italian development axis of Milano–Torino, and confidence must be inspired in Friuli becoming industrialized in a modern way and receiving growth industries; it must emerge from the shadow of Upper Italy. This is why

Table 3.18 School level and readiness to accept a job at home.

School level attained	Readiness to accept equivalent job at home		
	Immediately	Probably	No
primary school	51.4	48.0	36.4
middle school	34.3	28.0	9.1
high school, university	14.3	24.0	54.5

Table 3.19 Mastery of language and readiness to accept a job at home.

Mastery of foreign language	Readiness to accept equivalent job at home		
	Immediately	Probably	No
fluent	39.4	39.4	21.2
good	50.0	43.5	6.5
mediocre	58.5	39.0	2.4
poor	83.3	8.3	8.3

Table 3.20 Occupation and readiness to accept a job at home.

Occupation	Readiness to accept equivalent job at home	
	Immediately/probably	*No*
major building trade	100.0	0
supplementary building trade	100.0	0
mechanic	100.0	0
metal worker	90.9	9.1
no trade skills	89.5	10.5
service and administration	82.4	17.6
other crafts	77.8	22.2

there is concern for a regional university in Udine, connections with a trans-Alpine highway network and attempts to create a regional autonomist party in Trieste and in Friuli, interpreted as the gateway to Middle Europe and not just a northeastern appendage of Italy.

The growth impulse released by reparation of the damages of the catastrophe (Sec. 3.2) can lead to Friuli losing its previous backwardness, and acquiring an attraction for its better-educated emigrants as well. It must be assumed, however, that the latter, somewhat more demanding, returnees from places of emigration will incline more to types D and E (cf. Sec. 3.3) than those of types A and B of the settlement pattern (p. 000).

Notes

1 Dobler, R. 1980. Regionale Entwicklungschancen nach einer Katastrophe. Ein Beitrag zur Regionalplanung des Friaul. *Münchener Geographische Hefte* **45**. Regensburg: Kallmünz.
2 Burton, I., R. W. Kates and G. F. White 1978. *The environment as hazard*. New York: Oxford University Press.
3 Dacy, D. G. and H. Kunreuther 1969. *The economics of natural disasters: implications for federal policy*. New York.
4 Cochrane, H. C. 1974. *Predicting the economic impact of earthquakes*. Natural Hazards Res. Working Paper 25, 1–41. University of Toronto.
5 Dacy, D. G. and H. Kunreuther, op. cit., p. 8.
6 Dacy, D. G. and H. Kunreuther, op. cit., pp. 9–10.
7 Friesema, H. P. *et al.* 1979. *Community impact of natural disasters*. Beverly Hills: Sage.
8 Fritz, C. E. 1963. Disaster. In *Contemporary social problems*, R. K. Merton and R. Nisbet (eds), 651–94. London.
9 Fritz, C. E. op. cit., p. 692.
10 Quarantelli, E. L. and R. Dynes (1977). Response to social crisis and disaster. *Ann. Rev. Social.* **3**, 23–49, esp. p. 36.
   Friesema, H. P. *et al.* 1979 *Aftermath. Communities after natural disasters*. Beverly Hills: Sage.
   Wright, J. D. *et al.* 1979. *After the clean-up. Long-range effects of natural disasters*. Beverly Hills: Sage.
11 Herweijer, I. S. 1955. Het agrarisch herstel en de herverkavelingen in het rampgebied. *Tijdschift KNAG* **72**, 297–306.
12 Di Sopra, L. 1977. *Stima dei danni*. Documento sulla ricostrizione, no. 1, p. 40.
13 Stagl, R. 1980. *Planungen und Massnahmen nach einer Katastrophe und ihre Bewertung in der Friaul. Wiederaufbaustrategien und -probleme in Friaul*. Unpubl. master's thesis, Munich.
14 Wagner, U. 1979. *Wiederaufbau als Sanierungschance oder Fehlinvestition. Untersuchungen zur Mobilitätsbereitschaft in ausgewählten Abwanderungsgebieten des Friaul*. Unpubl. master's thesis, Munich.
15 Voelkl, H. 1980. *Die Rückkehrbereitschaft ausländischer Arbeitnehmer in ihre Heimatländer am Beispiel der Friauler in Deutschland und in der Schweiz*. Unpubl. master's thesis, Munich.

4 Perspectives on reconstruction

4.1 Initial alternatives

Relevant decisions to be taken during post-disaster reconstruction of an area with the historical depth of Friuli are more complicated than when it is a matter of rebuilding a city with a past of only a century or two. One layer, so to speak, of preordained decisions lies within the rubble of the disaster. A second layer was superimposed on this one when decisions had to be made about the emergency accommodation of the population, for the barrack cities not only tied up resources that had to be diverted from eventual permanent reconstruction; they also laid a burden on building plans for the permanent city, because the old one that lay in ruins, the temporary city (Italian: *baraccopolis*) and the future permanent one in many places criss-crossed and mutually impeded one another. Anything done to open up the barracks town means new streets, water systems, powerlines, sewers and so on, with what was intended just as a 'temporary' infrastructure network spreading out over the area of the commune and two or three conceptions of the city overlapping temporally, and not without some costs due to friction.

The questions of timing and form of reconstruction play a major role. Two points of view are conceivable in these regards, which were also subject to much disagreement after the rebuilding of cities destroyed by war.

(a) Should the 'identity' of a city that has developed over centuries be preserved by painstaking reconstruction, regarding the city as an historical heritage and trust, the re-establishment of which must take into consideration the traditional narrow street pattern even for the 'underground city' of the conduit systems for water, sewage, etc?

Or shall

(b) the rebuilt city amount in fact to a new structure, accommodating to new traffic patterns ('the automobile city') and more demanding standards for the supply of light, air and open space, thus becoming a more spacious open city (and in our case, safer from earthquakes)?

The reconstruction of war-ruined cities might be introduced here because it raises problems parallel to ours. It differs enormously in scale, however, from the considerations that have to be applied in an area only 4800 km² in size. After the war, it was mostly whole nations that took up reconstruction; interested groups were divided, labor was in short supply. The decisions about reconstruction in Friuli, however, are being undertaken in a time of peace, and they could be arrived at with the combined ideas of thousands of architects and planners in competitive discussion, and solutions could be reached with participation and aid from all over Europe. Even so, the first critics of Friuli's reconstruction have rendered devastating judgments.

Barbina[1], for example, considers that the ideas of the planners ran only in the narrow channels of emergency, with the goal of raising a provisional infrastructure and barrack housing from the ground, simply to avoid paralysis.

The provisional structures, he suggests, were not developed with an eye toward the eventual results. Their costs, he alleges, often wasted more than half the money needed for a real reconstruction. The barracks were alleged to contribute to the ruin of the land. Supposedly, they awakened false expectations as to which communes could be rescued permanently.

Friuli, Barbina argued, on May 6, 1976 lost not only its buildings but its history. Geologists and engineers could reconstruct the buildings. But to reconstruct the pattern of human activities as it existed in this area before would require that the expressions of a culture and a way of life be brought back in order to restore to the people their lost identity. Presumably, bits and

Figure 4.1 The Cathedral of Gemona, built between 1290 and 1337, was badly damaged and lost its right aisle. Its rosette, its gallery of sculptures and, above all, its huge statue of St Christopher belong among the most remarkable works of Gothic art in the Alps. A crack through the giant statue is said to date back to the Villach (Austria) earthquake of 1348. The great age of the monuments, intact through the ages until 1976, shows that devastating earthquakes happen here at intervals but are therefore all the more dangerous.

114

pieces of the whole made functionless during the course of history would not deserve reconstruction at all. Settlements on the Tagliamento that had made sense in the days of the rafting trade, places in the high mountain valleys founded on the sparse subsistence standards of the Middle Ages with no present economic basis, had lost their rationale. The same would apply to house form. The dispersed pre-Alpine farmhouses and closed farmyards no longer need provide ample storage and shelter, and might get along without all the space provided for families with numerous children and servants who no longer exist.

These considerations throw light on the important regional economic problem of reconstruction in Friuli from one side of the argument. Two solutions were conceivable:

(a) gradual reconstruction *in situ* in traditional styles, but more secure against earthquakes, or
(b) the importation of the missing housing stock in the form of prefabricated houses made elsewhere and merely assembled inside the disaster area.

Since 20 880 of such units were first planned for, the total disposable funds for which a choice had to be made, if each of the prefabs was worth about $5700, amounted to around $117 million.

Solution (a) would have generated a high internal income effect within the region at first for the local construction industry, reinforced by that of all of northern Italy. In some cases, it would even have been feasible to bring about a reversal of the emigration balance by providing employment at home for the many Friulians working abroad, and jobs could have been made available in Friuli, at least for some while. It is well known that many foreign workers are employed in construction abroad. But the local rural population possesses skills in building and repair, even if only on a handicraft level. If they had been given the amount of money spent elsewhere by the Italian government and foreign relief organizations for imported prefabricated housing, it would have been possible to achieve more through the organization of self- and neighborhood help with the guidance of experts, in the form of a sort of unemployment relief in some cases, than was achieved by building barrack towns. The identity especially of the smaller rural settlements would be ensured; money would mainly remain at home; the reconstruction process would be spread out over several years. This idea, which the architect Roberto Pirzio Biroli particularly stood for, can be described as follows.

The inhabitants of a destroyed or uninhabitable house gradually move back from earthquake-proof 'living cells' designed by experts (that is, from the minimal accommodation also offered in the form of camping trailers, garages, railroad cars and tents to those who did not care to move out after the May earthquake) into likewise earthquake-proof building frames according to the progress of construction. They gradually take possession of space in this fashion. Evacuation would have been almost unnecessary, and the buying power of the people subsidized in coastal cities would have stayed in the home communes. Finally, it should be mentioned that original farmhouses and town houses of the medieval city centers had probably not been built in a few months either, but had come out of a process of additions and repairs lasting for generations.

Solution (*b*) produces the largest effect on income in the areas where 'prefabs' are built (for example in Austria, West Germany, Yugoslavia, Switzerland, Canada and other parts of Italy). Established firms received an impetus, and new ones were on occasion founded without much previous preparation. The 'teething problems' of their products were tried out first on the Friulians. The local construction industry, however, supplied only foundations, connections to utilities, and interior finish. Mobile assembly teams of the prefabrication firms spent only part of their wages in the disaster area, so that multiplier effects from this source were very modest. Building activity was compressed into a few months by putting up whole quarters of prefabs.

Certainly, space is thus created rapidly, but it is of dubious quality; it is 'housing' more than homes. The personal relationship of the residents to the standard mass-production houses, crammed tightly together on someone else's appropriated land, may have been much looser than those that might have developed if the means devoted to the prefabs had been used to repair the family house on the hereditary sites. From the standpoint of population policy, the holding power of the prefabs in the long run had to be inferior to that of the old houses restored by Biroli's method. Their earthquake resistance, on the other hand, is surely as good or better. They also made it possible simply to provide accommodation quickly, and thereby to release the Adriatic cities when the tourist season began, which was one of the conditions when evacuation began in September. Our investigation must extend as well, accordingly, to the problem of the perception of the risk connected with remaining in old or new buildings, to long-term effects and to the impact of various imponderables of demographic development in the area.

First, however, let us pose the issue of the reconstruction in the form of prefab housing in its quantitative dimensions, spatial differentiation, and the approach of the affected population to it. The criticisms that will arise from this, and the ones just mentioned, should not, however, be misunderstood as a general rejection of all planning. The strategy proposed by Pirzio Biroli as an alternative was not available in the beginning but only resulted from a learning process. It is debatable whether the example of Sta. Margherita del Gruagno (Commune of Moruzzo) could be applied as a paradigm to any and all communes. This is a matter, among other things, of regional manpower and citizen participation. It would not be possible simply to duplicate the special case of an architect and conservationist such as Pirzio Biroli – at home in the region and owning property there, whose charismatic appeal, ability to convince and bureaucratic adroitness opened the way for people themselves to have a say in the control of reconstruction and to refuse to accept the 'barracks' – who provided funds, and fought the matter through against reluctant officials.

For one thing, there are not enough architects properly trained for such jobs. The latter are mostly prepared by their education for building new buildings, and less for the restoration of old ones. This may change when, under the pressure of development costs, there will be a transfer in Europe in housing creation from new construction to the renovation of old structures, but of course it cannot be assumed as part of the training of the architects available at present. Therefore the 'giant building project' of Friuli has a wider relevance and so will be watched closely by professional engineering groups in neighboring countries.[2]

116

The extent of the reconstruction program can be indicated by some figures. The San Francisco earthquake of 1906, with the massive fire that followed, wiped out 28 130 houses. The building program of the Commissario Straordinario originally contemplated 20 880 houses. If we take four persons as an average for a family, the number of prefabs roughly corresponds to the total of approximately 80 000 homeless in the two earthquakes (cf. Fig. 1.6).

Planning for the prefabs in 96 communes was undertaken on the basis of Law 227, May 13, 1976, which was amplified by Laws 336 and 730 after the experience of the September earthquake. The program was declared officially terminated on April 30, 1977. The law coordinated:

(a) the plans of the Commissario Straordinario for 9991 dwelling units totalling 406 298 m^2;

(b) the plans of the Autonomous Region of Friuli–Venezia Giulia with 9281 dwelling units making 343 000 m^2; and

(c) the construction activity of public and private relief organizations, coordinated by the Commissario Straordinario, with 1250 dwelling units and 52 000 m^2

Altogether, 20 992 dwelling units with 813 298 m^2 were produced, giving shelter to more than 65 000 persons. Over a period of eight months, an average of 10 000 m^3 daily and 300 000 m^3 monthly of rebuilt space were created. This amounted to 12 m^2 or 36 m^3 for each homeless person. However, this volume represents only a fraction of the living space available to the Friulians before the earthquakes. Table 4.1[3] gives an overview of the types of units employed. Prices per square meter fluctuated, according to what was published, between 98 000 and 120 000 lire per m^2, ($130–$160), so that with approximately 12 m^2 per resident about $1600–$1900, or an average of $1750 per person were spent. The construction program therefore cost around $120 million.

Table 4.1 shows what kinds of houses were set up in the zone of the catastrophe. They varied in quality, and the homeless people naturally were concerned to be placed in the better house types. Representatives of prefab firms also tried to exert influence on the ordering by the communes. Not all the house types were suitable for mountain areas. Foreign donors, for whom Friuli figured in the mental map as part of 'La Bella Italia', overlooked long mountain winters and heavy snow falls, as occurred in the harsh winter of 1976–7. A special little industry arose (cf. Figs. 1.13, 2.14 & 4.2) devoted to correcting the worst defects of construction, for instance, building extra roofs over the Canadian containers.

Since the buildings begun earlier and finished sooner showed defects that could be corrected in later programs, the evacuated population hesitated at times in accepting certain housing types. Disappointments can result, particularly for foreign relief organizations, over the 'ingratitude' of those on whose behalf so much helpfulness and generosity have been mobilized. Mitchell[4] lists, as reasons for such a refusal of aid in the Turkish disaster area of Gediz, opposition because of disagreement over the choice of locations for new buildings, the quality of construction, because of the size of the houses (which did not have space for domestic animals), and – in an Islamic area – because people could see into the homes through windows set too low. He also criti-

Table 4.1 Types of prefabricated houses.

Building firm	Region	Emergency Commissioner	Communes	Total surface (m²%)	
(1) Volani	100 115	—	—	100 115	13.72
(2) P. Della Valentina	66 223	—	13 461	79 684	10.92
(3) Krivaja	—	—	61 450	61 450	8.42
(4) Pittini	42 838	—	—	42 838	5.87
(5) Tecna	39 856	—	—	39 856	5.46
(6) Atco	—	37 210	—	37 210	5.10
(7) Sicel	35 095	—	—	35 095	4.81
(8) Ars et Labor	—	—	26 408	26 408	3.62
(9) Morteo Soprefin	3 191	22 547	—	25 738	3.53
(10) Cocel	—	—	25 000	25 000	3.43
(11) Tacchino	19 059	—	5 809	24 868	3.41
(12) Trybo	—	—	23 925	23 925	3.28
(13) Pasotti-Notari	—	—	23 726	23 726	3.25
(14) Industrie Carniche	10 944	—	7 282	18 226	2.50
(15) Socomet	—	—	16 841	16 841	2.31
(16) Bortolaso	14 594	—	—	14 594	2.00
(17) Meccanocar	—	—	14 176	14 176	1.94
(18) Commerciale Tecnica	—	—	13 371	13 371	1.83
(19) Superkonstruksjon	—	11 908	—	11 908	1.63
(20) Fina	—	—	9 012	9 012	1.23
(21) Cemi	—	8 928	—	8 928	1.22
(22) IPM	—	—	8 800	8 800	1.21
(23) ICM	—	7 560	—	7 560	1.04
(24) Precasa	—	—	7 000	7 000	0.96
(25) Officine Ceccoli	—	—	6 500	6 500	0.89
(26) Tuscania	6 018	—	—	6 018	0.82
(27) Presmont Vega	5 364	—	—	5 364	0.74
(28) Coraf	—	—	5 163	5 163	0.71
(29) Edil Morena	—	—	5 030	5 030	0.69
(30) D. F. Allessandrina	—	—	4 151	4 151	0.57
(31) Adua Case	—	—	4 800	4 800	0.53
(32) Intercamp	—	3 780	—	3 780	0.52
(33) Habitat	—	—	3 200	3 200	0.44
(34) Benigni Bononi	—	—	2 803	2 803	0.38
(35) Caravan	—	2 250	—	2 250	0.31
(36) Sermet	—	—	1 115	1 115	0.15
total	343 298	94 183	292 116	729 597	100

Source: Ricostruire, Revista Tecnica di Informazione **1** (1), 7.

cizes the lack of participation on the part of those affected and the absence of alternatives from which to choose.

Mitchell's example ('. . . thirteen houses in a row parallel to and twenty-five feet from a dirt road with the closest water three-quarters of a mile away . . .')[5] has parallels in Friuli; for example, in respect to location along railroad lines and heavily traveled roads. Here, the planners obviously have overlooked the fact that the nightly passage of trains past the flimsy prefabs would give rise to similar conditions to those of the earthquakes in the shaking and the noise they made, and did not take seriously enough the traumatic

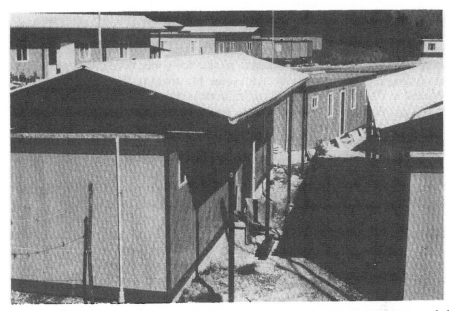

Figure 4.2 Already, during the first winter, many of the containers started to leak, because their roofs could not carry the pressure of snow. Since the containers were like saunas in summer and like damp refrigerators in winter, some of the occupants helped themselves by adding small roofs. Through the thin walls, all neighbors can involuntarily participate in family life. In the long run, crash housing of this type must lead to emotional problems.

experiences of their potential occupants. Some ensuing sections will be dedicated to the consideration of such elements of the residential habitat.

One of Mitchell's key conclusions can be taken as a precedent for Pirzio Biroli's hypothesis: '. . . a smaller number of inhabited houses is more beneficial than a larger number of uninhabited ones. Funds spent for houses that may never be occupied could be more wisely distributed for repairing damaged houses . . . Close coordination and understanding between government planners and villagers could alleviate most of the mutual dissatisfactions. . . .'.[6,7]

The problem of reconstruction after a catastrophe cannot simply be solved technocratically, but requires considerable social-scientific verification of all measures adopted. Consequently, not only engineering damage estimators, geologists and housing planners should be part of the 'zero hour' team, but also social psychologists and regional planners acquainted with empirical social-science methods. Buildings are not an end in themselves, but are built for people, and it is one of the most prominent tasks of those whose job it is to spell out the results of catastrophes to find out about people's probable reactions to them. Regarding such reactions, it must be taken into account that the hasty assignment of major responsibilities in more or less chaotic conditions subjects the integrity of decision-makers to a severe test of sturdiness. Prosecutions in the summer of 1977 showed that not all of them could resist the temptations of corruption.

119

4.2 Implications of the locational assignment of prefab quarters

Even if the city rebuilt is supposed to be the aim of reconstruction, people are compelled beforehand, nonetheless, to live for some time in the prefab quarters (some say it may last a decade, some for always). These so-called 'baraccopolis' settlements exert decisive influences on the perceptions and behavior of the people established in them. The planners of such settlements, consciously or unconsciously drawing on certain thematic images or ideologies, influence the thinking of the residents about their living conditions. For this reason, a number of such implications will be brought out in this chapter. One of the first choices to be faced, for instance, is whether to place the barracks on the original property sites of former house owners or clustered together in a special tract. Colloredo di Monte Albano provides an example.

Approximately 100 'prefabbricati' have been erected wherever possible on the individual sites of the residents and can therefore continue to use existing service facilities (streets, sewage, water supply, electricity). The owner thereby maintains close contact with his former farm buildings, immediately restoring his stables. But this prevents any reduction in density of the former too densely built-up area of the old village. Osoppo, on the other hand, has plans for co-operative livestock raising in the outer portions of the community fields.

Contact with the old building is advantageous for an efficient use of time. In contrast, a disadvantage is that individuals' construction work is fragmented and joint use of building equipment ('cantiere simultanea', see below) is impossible. The groups in the village best able to invest are able to proceed faster with reconstruction than are poorer folk. The form of the village accordingly becomes disunified (with negative effects on tourism – Colloredo Castle was the symbol of Friuli). Individualized and fragmented reconstruction on single sites prevented any spirit of unity from developing and precluded land-use planning resting on a general consensus among individual decision-makers. The commune has, likewise, no chance to lay its hands on sources of financing that make larger resources available for planning above the individual level. In this commune, everything remains as it was. (It is also plagued by disparities between the reconstruction plans of the owner of the castle and the commune.)

For our next example, midway between individualistic fragmentation and the strict planning of a baraccopolis, let us take Magnano in Riviera, with its approximately 2000 inhabitants (Fig. 4.3). Here, the prefab quarters were set up in groups of from 36 to 52 houses at five different spots in outlying parts of the commune. This partly took advantage of an existing subdivision of land already allocated for building. Four adjoining houses were erected on each site (Fig. 4.4). One of these, if possible, was set aside for the owner of the site. As emergency indemnification, he received a lease amounting to an eighth of the value of the land before the catastrophe. It is intended, as reconstruction in the center of the commune gradually progresses, to have the occupants vacate these four-house complexes on the outskirts, as the living quarters of the people coming back into the center are surrendered to those who have been staying in the outside zones. Ideally, the owner of the site would then stay behind as the last occupant of the consolidated prefabs, and after they were taken away could start on the construction of his own house. The

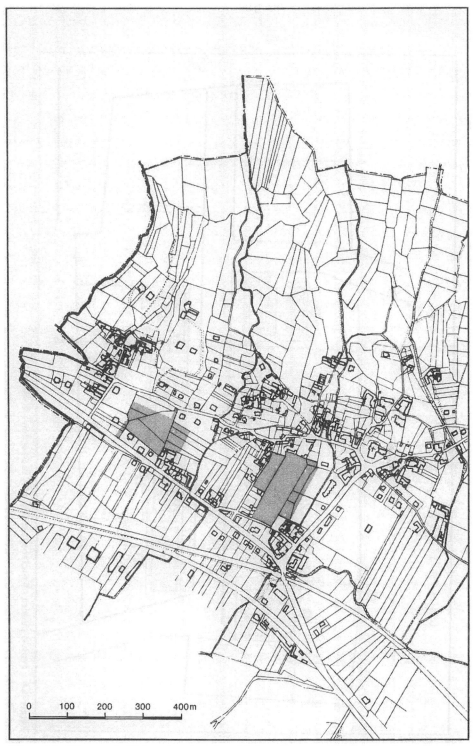

Figure 4.3 Magnano in Riviera: location of prefabricated houses within land ready for development.

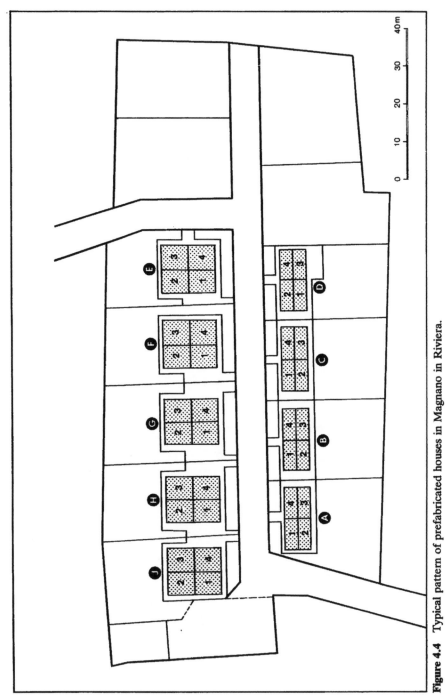

Figure 4.4 Typical pattern of prefabricated houses in Magnano in Riviera.

investment in services for the emergency accommodations in the subdivision land in the outlying area would thereby not be forfeited.

The examples thus far cited come from the hill country. It becomes more complicated as we come nearer to the mountains and space suitable for erecting prefab quarters decreases, or when we are dealing with cities of special historic value such as Venzone.

Venzone, with 2652 inhabitants, 49 dead, and around 60 per cent homeless, was one of the hardest hit communes. Because of the high standing of its historic inner city, here, as in the case of Gemona or Osoppo, there were reconstruction problems concerned with the protection of art treasures. Their restrictive effect had tended to freeze economic activity in Venzone before the earthquake, since any change of land use or intensification of use was hedged about with many restrictions. The earthquake, totally wiping out the existing structures, opened the way for renewing the center of the historic city more in accord with its functions. At the same time, Venzone can be taken as an example of the limitations that are imposed because of hazards when reconstruction is undertaken:

(a) If planning estimates evaluate the risk of landslides as particularly serious (especially in the Portis section of the commune), then new land must be sought at an appropriate distance from the threatening steep cliffs.
(b) This level land away from steep slopes can only be found in the vicinity of the Tagliamento river bed, which presents a serious risk of flooding. Between the Scylla of landslide and the Charybdis of flood only the existing site of Venzone, evaluated since Roman times as most advantageous, presents a reasonable location.
(c) Nevertheless, the fault lines that come together in its market place attest to the highest earthquake risk.

We shall deal with these hazards in Section 4.4. As to the location of the prefab quarters, the population of the destroyed city protested vigorously against an area south of Venzone, used as a camping ground for a tent settlement, as the place for setting up the prefabs (flood danger, high winds), even though considerable investments had been made at this location. Therefore, after the outlining of this plan before a meeting of the citizens in Lignano, the communal government had only the option of putting the temporary quarters in such places where:

(a) the reconstruction plans would not be interfered with;
(b) there was access to the city center, making feasible the later use of existing infrastructure; and
(c) concentration was as dense as possible, in order not to have to place too great demands on agricultural land.

Therefore, 12 individual housing blocks with 822 apartments were established on communal or confiscated private land around the ruined part of the inner city exempted from the historic preservation regulations (Fig. 4.5). The government of the commune counts on an occupancy of five years at least for these accommodations. It seems hardly conceivable that the most exposed, at

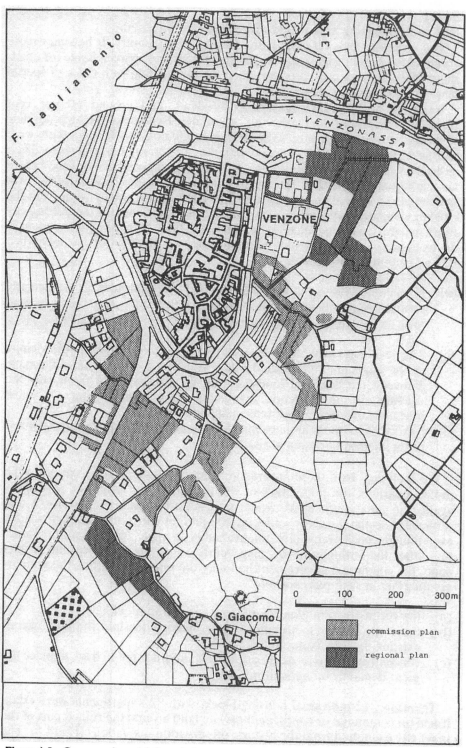

Figure 4.5 Groups of prefabricated houses surround the destroyed center of the town (Venzone).

least, of these rows of prefabs laid out along a busy through highway can be expected to have occupants for such a period. There are indications, anyway, that the selection of at least one of the two most exposed areas is intended to produce a 'demonstration effect'. The prefabs used belong to four types:

Meccanocar (Toscana) at 140 000 lire/m^2 = \$164
Volani (Trentino)   at 120 000 lire/m^2 = \$140
Della Valentina (local) at 120 000 lire/m^2 = \$140
Tecna (Piemont)   at 120 000 lire/m^2 = \$140

They were provided free to the homeless according to family size, time of the request, and residence in Venzone, with two different floor-space variants.

Meanwhile, the historic center subsidized by contributions from foreign relief organizations, sponsoring towns, cultural organizations from abroad, and so on, and with the support of former business interest, has been cleared, and is supposed to be rebuilt according to the plans of R. Pirzio Biroli.

After experimenting in a much smaller commune (Santa Margherita dal Gruagno, a section of Moruzzo) this architect established the model of a 'cantiere simultaneo' or sort of construction workshop, into which the people affected put their compensation money and personal savings for concurrent reconstruction and restoration. Through multiple use of construction equipment (e.g. erection of a crane for a whole block of individual houses) the costs

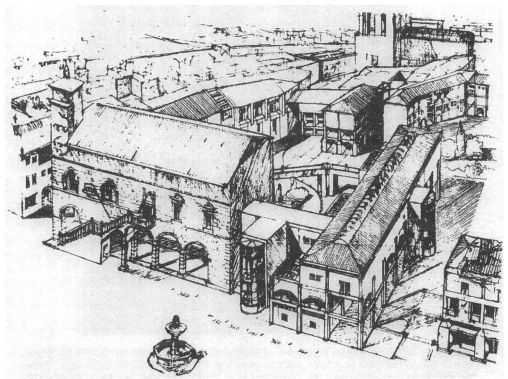

Figure 4.6 Modular elements used for the reconstruction of Venzone (source: *Ricostruire*, monografie 3, p. 5. Udine, 1977).

can be kept lower than normal. In Venzone's reconstruction, an attempt will be made to base the reconstruction on module-elements derived from units and multiples of the classic Friulian measure of 7 m, a unit in turn determined by the length of logs used in rafts on the Tagliamento (Fig. 4.6). When such modular elements were assembled, experience with earthquake-proof methods of assembly was applied and the reconstruction of Venzone is supposed to combine facade elements thus rescued from the debris and modern elements in a way that reflects the city's character but increases its functional qualities.

During the discussion of reconstruction plans an interesting issue arose. As an example of a medieval city of rare completeness, Venzone enjoyed a

Figure 4.7 Venzone, with its population of 2600, dominated the medieval trade in the Tagliamento Valley. Enclosed by a huge wall, the town was a jewel of medieval architecture. When the four town gates collapsed, the inhabitants were caught in a trap. So, for some time they demanded to fill the moat with the debris of the wall before they would agree to return to the city center.

126

certain income from tourism, for it is situated directly on the approach route for all holiday-makers coming from the North to the Adriatic (Fig. 4.7). The old town was surrounded by a high wall, Venzone's especial pride, and strictly protected as an ancient monument, which meant that alterations to the buildings and investment in new developments were minimal. The total destruction of the town made it possible to retain the old facades and the classical proportions while endowing the reconstructed town with a more rational groundplan and modern hygienic facilities, thus handing it back to the inhabitants after a thorough clearing-up of the confused patchwork of property rights, which sometimes included shared ownership.

Although the commune can bring a wide range of pressures to bear to encourage consensus among its citizens, and can even undertake compulsory purchase if owners drag their feet, many proprietors were for a long time (until 1979) reluctant to return to the city core. There was a real danger that the first buildings to be reconstructed would be public ones, financed by charities and state funds, while the citizens themselves, relying on private investment, failed to follow suit. Further inquiry revealed that psychological trauma played an important part in this reluctance.

The city wall of Venzone, pierced by only four fortified gateways, which formerly had been the pride of its citizens and of not inconsiderable economic value as a tourist attraction, acquired in the course of the earthquake a new and negative significance. It was now seen as a fatal trap. The four gateways of the city were all blocked with debris after the earthquake. The population, for the most part advanced in years, was unable to escape from the closely built streets of the old town. Most of the 49 of Venzone's dead died beneath the ruins of the medieval city. Hence, the demand was widely expressed to fill in the city moat with the material of the wall which stood behind it. In the stress of the disaster the symbol and token of the city of Venzone underwent a change in value from positive to negative. But it will be reconstructed.

In this sequence of examples, the rebuilding of an old town outward from an adjacent prefab center ('baraccopolis') can be demonstrated best by Osoppo, with 2543 inhabitants before the earthquake, 104 dead, and 759 totally destroyed dwellings. The city is a key industrial commune furnishing about 2000 jobs, with plenty of land to build on where the Tagliamento leaves the mountains. Unlike Venzone, a communally owned, contiguous area, augmented by a few condemned sites, on the margin of the almost completely destroyed urban center (fully vacated by the end of March 1977) could be used, therefore, after a few relocations of the preceding tent city. The commune lies at the foot of a major fortress whose steep cliffs pose a landslide threat. In Figure 4.8, a portion of the defensive works at the foot of the fortress-mountain between prefab blocks R5–R4–R6 and R3–R1–R2 is still recognizable.

Within the Rivoli section of the commune, southward from Osoppo, is one of the most concentrated industrial districts of Friuli. The commune population of Osoppo therefore exhibits a decidedly positive attitude toward reconstruction in terms of jobs and also on the basis of behavioral imponderables.

The adjacent industrial zone, the influence of which extends widely over the surrounding area, attracts commuters from both the mountains and the hill country (cf. the favorable standing of Osoppo by most indicators that were applied in thematic mapping or correlation analysis.)

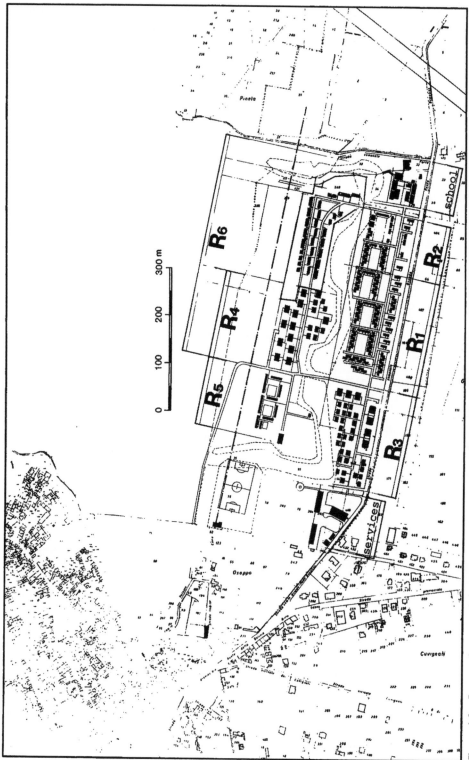

Figure 4.8 Osoppo: location of the 'baraccopolis' at the outskirts of the destroyed core of the city. Note 'corte' at R1 and R2.

Unlike Venzone, where architectural planning marked by R. Pirzio Biroli took precedence, for Osoppo at the time of the research there still were no clear formal urban concepts concerning reconstruction under the impetus of an advocacy planner, but there perhaps existed definite community ideas about how to go about reconstruction with greater citizen participation. Its major features are revealed in the plans in Figure 4.8 (see also Fig. 4.9).

In the prefab town, the prefabs, partly taken over from the Province of Tusca where they had once been used in a castastrophe, were erected in the form of a court. Each of these dwelling-courtyards corresponds to a residential cell of the old town, and each courtyard is represented by four delegates before the 'ufficio tecnico' and city government. Every Friday evening there are open sessions of the planning experts with the delegates. Planning is surely not made easier by this, but stays closer to the citizens.

Osoppo had possessed a land-use plan before the disaster, which was consulted in the reconstruction planning, with its service-axis of public facilities (school, sports facilities, etc.) The delegates of the individual courtyards, who represented their fellow citizens in the building commission of Osoppo, spoke up against the center for the elderly that had been planned within this axis of services, outside the destroyed town center. Alternatively, it was proposed to carry over the demographic structure already present in the prefab courts ('corti') into the reconstructed units of the old town as planning occurred.

Figure 4.9 Together with Figure 4.8, the picture illustrates a typical 'baraccopolis' (barracks town): a monotonous lining-up of different types of prefabs of different qualities. In the foreground, the walls of an old fortification and three houses of a better type with two storeys can be recognized. Behind the baraccopolis, an old military airstrip shows an assembly of railway cars in which many inhabitants of Osoppo spent the winter of 1976–7. In the background, the ruins of Gemona cover the alluvial fan. The dark rectangle in the far left corner of the fan's triangle indicates the Central Hospital before dynamiting.

Table 4.2 Prefab quarters of Osoppo.

Subdivision + plan	Area (m²)	Type of prefab	Number of houses for				Total	Families	Persons	Area (m²)
			1 family	2 families	3 families	4 families				
R 1 + 2 r	45 600	Tuscania	136	—	—	—	136	136	402	3 196
R 3 r	19 380	Tecna	4	31	1	—	36	71	161	3 090
R 4 r	16 800	Tecna	5	8	5	4	22	53	180	2 712
R 5 o	8 800	Finsider	12	6	—	—	18	24	66	648
R 6 c	24 305	Krivaja	50	—	—	—	50	50	179	2 448
total	114 885		207	45	6	4	262	334	988	12 094

Plan: r, regionale; c, commissariale; o, others.

Although architectural form had not yet been considered, it is already plain that the courtyards of the old town that was to be rebuilt should be more healthy, more open, have more green space, and should be functionally distinctive as would befit a rural area with community centers planned as a counterweight to the industrial zone of Rivoli (Osoppo). The subsidiary agricultural enterprise could also be put on a sounder economic footing by these means. Thus, there is a multiphased planning effort in Osoppo which seeks to solve more than just one of the problems at hand.

The system of courtyards, which requires planning close to citizen opinion, cannot, however, be instituted in the same way everywhere, because of the variety of different housing types and the overlapping of plans (piano regionale/piano commissariale). Table 4.2 summarizes the main data, and shows what problems can arise just because of different housing types. In the settlement for nearly 1000 occupants shown in Figure 4.8, there is an average of 12.24 m^2 per person available. The Tuscana-type prefabs set up in areas R1 and R2 were already in use and only provided 7.95 m^2 per person, while the Finsider-type was set up in such a way at the time of the research that each resident had 9.82 m^2. The Krivaja-type (13.67 m^2) in area R6, Tecna (15 m^2) in area R4, and Tecna (19.2 m^2) in area R3, on the other hand, provided more than average space per person. Jealousy and accusations of favoritsm could be expected. In addition, 756 people lived in all sorts of dwelling types on their own sites and thus took no part in the process of opinion-making in the main settlement of the 'baraccopolis' or in the community life developing there.

It should be clear from what has been said that Osoppo represents an experimental situation in town building, in which very diverse grand themes of community policy are realized. The success or failure of such variants of reconstruction can be a lesson for future disasters.

4.3 Effects of external intervention

How much should external relief organizations be allowed to influence reconstruction policy and planning? From Section 4.2, it was plain how much political ideas can enter even into apparently inconsequential details of the temporary accommodation of the homeless in barracks, but still much more into the ultimate reconstruction of the city. Thus, the planning units in Osoppo just described were undoubtedly influenced by the ideas of left-wing planners, probably from Bologna. On the other side, conservative circles sought to turn aid from foreign charitable and religious relief institutions to the support of their own concepts.

The charitable organizations from countries providing relief are confronted here with the task of taking on the function of 'stabilizing the system' by underwriting individual houses on their own sites, while elsewhere left-orientated communal governments are trying to achieve a decrease in the bourgeois character of the population by means of large collective buildings. The 'laboratory conditions' in Friuli make it possible to trace out differing planning ideas according to political power relationships. Results of an inquiry made among some of these relief organizations will be reported later.

A material symbol of the internal tensions within communes is found in the controversy over the rebuilding of the churches. In communes that have anti-clerical majorities, there was sometimes the tendency to have done 'in one sweep' with the church as the ruins were cleared away. This 'saved' long discussions about what was worth saving from a cultural–historical standpoint and what could be saved from an engineering standpoint. Mayors who decreed the abolition of church ruins could argue from the basis of economy that whatever Italian, Canadian or German military engineers had cleared away at no cost in the first clear-up operations would make superfluous any later rubble clearing jobs let out at high prices to private firms.

If this worked out, and the material symbol of the 'old order' was abolished, redevelopment plans could start out from a complete *tabula rasa* and the priority list would look different than it might if 'rebuilding of the church' was there as a requirement returning on the agenda of every meeting of the city council and interfering with more pressing problems. The secularization that had taken place between the original construction and the reconstruction of such a church in a particular commune was evident in the fact that reconstruction often has to be financed by the church authorities themselves and is no longer even fought for by the local faithful.

The Catholic church was foresighted enough, in view of the need in which the population found itself, not to press for the collection of funds for rebuilding of the churches. On the contrary; it spread the motto, 'Prime i case, dopo i chiese' (houses first, churches afterwards).

Foreign relief agencies did not always have a clear idea at first of the ethnic composition of the population. Mostly not knowing the language, they took all the recipients of their aid for 'Italians'. Only people from the immediately adjacent countries – Austrians and Yugoslavs – were precisely aware of ethnic relations, and sometimes helped the groups most akin to themselves – the Yugoslavs the Slovenes on the border, the Austrians the Germans of the Canale Valley and small isolated communes like the German-speaking Sauris. It was also not always clear to foreign relief organizations that the provision of barracks, as the local people saw it, ought to fit the local needs and not provide a 'surplus' which would attract outsiders to the place. The barracks should definitely disappear as soon as houses were rebuilt. Otherwise newcomers from the Italian Mezzogiorno might be able to move into the houses thus released, introducing the danger of having too many outsiders in Friuli, with a swamping of the native population by strangers and a shift in political majorities.

Whereas, on the one hand, certain elements of the population feared that an excessive supply of prefabs might attract newcomers from the South, other groups had the opportunity, under the same conditions but in a different demographic situation, namely in the mountain communes far removed from avenues of communication, of winning back members of the Friulian ethnic community who had emigrated by means of the availability of the prefabs. The slow process of emigration, such as can be documented in the mountain community of Lusevera,[8] was at least toned down by the contributions foreign relief organizations made for reconstruction. Thus the World Lutheran Fellowship presented 10 prefabs to the Cesariis di Sopra section of Lusevera commune (buildings of the Haas-model from Fellbach in Württemberg) which canceled plans already discussed in the commune for a full transfer of the population away from Cesariis. This kind of prefab, in contrast to the Canadian containers ('Atco') from the program of the Commissario Straordinario set up in many communes, represents a particularly solid kind of wooden house able to stand up to the more than 3000 mm precipitation in the commune. It may have meant an improvement of living conditions as they had existed before the earthquake. Here, as in the case of Sta. Margherita del Gruagno, it was one person who possessed a special gift for persuasion. Pastor Lindenmeyer of the Evangelical parish in Bozen, who carried over a strong motivation for the preservation of Cesariis di Sopra from the parallel situation of the South Tyrol mountain communes. He was able to call on the resources of the Lutheran Fellowship for preservation of this commune section, even though it has no services or non-agricultural employment base and can only be reached over a narrow, badly eroded mountain road.

So foreign relief organizations came in chiefly where, in their opinion, the Italian authorities delayed too long with their aid, perhaps because they had not decided whether a commune fraction was to get barracks at all. The foreign relief agencies to an extent participated in this decision-vacuum of 'Let's not say no just yet'. It is altogether understandable in human terms that foreign aid personnel should take such hesitation on the part of the responsible authorities for incompetence and spring into the void they saw for humane motives. In order to learn more about this, we interviewed nine international organizations in June 1977.[9]

The four-page questionnaire was filled out in detail by experts in these bodies and provided information on the following points among others.

Question 7. On whose initiative was your engagement in Friuli undertaken?

Question 9. What sort of measures were employed?

Question 10. In what communes did you invest, in what, and on what scale?

Question 13. What considerations induced you to concentrate on these particular communes? Could you kindly cite some of the important decision criteria that led to selecting these communes?

Question 14. Did other organizations, agencies, or persons support you in these decisions? If so, which ones?

Question 15. On the basis of what you know now, would you consider your decisions at the time to locate in those communes correct, and why?

Question 16. With what official agencies did you co-operate?

Question 17. How would you characterize this joint effort in Friuli?

Because of the confidentiality of information from these relief organizations, the answers to the questions cannot be reported directly. They were, for this reason, summarized in matrix form (Table 4.3). The matrix reveals that beside immediate help such as clothing, consumer goods and money that was spatially not very significant, a large portion of the approximately $11 million contributed by these nine organizations consulted was largely invested in prefabs and infrastructural facilities, such as schools, kindergartens, and dispensaries, in such a fashion that long-term consequences for the mobility behavior of the population may be expected. Unlike the unloved containers furnished officially by the government and subject to recall, for instance, in the case of contributions from relief organizations substantial and permanent gifts to individuals and communes were concerned, and not equipment lent only for emergency accommodation.

Location policy took account of small, less well known valleys and remote

Table 4.3 Statements of Austrian, German and Swiss relief organizations.

Index no. of organization	Type of activity			Scale of activity			Location			Criteria of choice of location				Assessment of locational decision					Assessment quality of co-operation				
	Emergency aid	Prefabs	Infrastructure	Low	Medium	High	Large communes	Small, less known communes in the valley	Remote mountain communes	Accepting decisions from other authorities	Known size of damage	Willingness of communes to co-operate	Need of most helpless communes 'neglected' by the Italian government	Right (no comment)	Right (great need)	Right (great gratitude)	Right (success in activating inhabitants)	Bad experiences too	Very positive	Positive	Not always positive	Good with communes	Less good with province
1	x	x	x			x	x	x	x	x												x	
2	x	x		x			x	x	x			x	x	x	x								x
3	x	x	x			x		x	x			x	x				x					x	
4	x	x	x			x	x	x	x			x	x						x			x	
5	x	x		x			x	x	x	x				x	x								x
6	x	x			x			x	x	x					x						x		
7	x				x		x	x				x		x	x	x					x		
8		x			x			x				x	x	x	x						x		
9	x	x	x			x			x			x		x	x							x	

mountain communes as well as of the central catastrophe zone on which the media kept concentrating in the earthquake period, and into which the Italian government put its chief aid efforts as well.

Among criteria for selection, the high degree of destruction brought relief activities primarily into the large valley communes near the epicenter. But the desire to help the socially underprivileged quite understandably and clearly expressed by all charitable and religious aid bodies, and the 'neglect on the part of Italian officials' that half the agencies reporting alleged to a considerable extent, led to their giving help to small and remote places. This compensated for the conscious or unconscious underestimation of the need for aid, primarily in the northern mountain part of the disaster area and near the border.

As well as the motives mentioned, ecological ones also played a surprising part among criteria for locational choice. A letter published in a newspaper[10] brings this out especially well. In this letter, the person who had organized help for Cesariis di Sopra in the commune of Lusevera (CML 41) takes exception to an earlier article in the same paper.[11]

Finally on 23 March ten houses from the firm of Barth could be left to their fate in the mountain peasant village of Cesariis near Tarcento. The Evangelical community of Bozen (Bolzano), along with friends and numerous anonymous donors had been collecting funds for these houses for a long time and had presented the money thus received to the diocesan charity agency with the request that they take care of the rest of the costs. The government did not want to rebuild Cesariis, but it had meanwhile set up accommodations there itself that were by no means as good and sturdy as the "Evangelical houses" that we are contributing.

Why did we insist on rebuilding Cesariis? The government sits in Rome, far away. Mountain peasants' problems in other provinces as well are often still heeded either not at all or all too little. On the other hand, a model in this matter is provided by the South Tyrol government.[12] For here in the mountains we know that the depopulation of the mountain districts means catastrophes in the valley. The Belluno-Longarone disaster can be traced back the same way: The honeycombed mountain, being no longer exploited, broke away and plunged into the reservoir, destroying the dam.

Mountain people have a special attachment to their home. The land handed down and tilled for centuries still means something there. People there still know how to deal with country streams and use pastures in such a way that the soil will not give way and crumble. Every peasant who stays in Cesariis (or some other mountain village) contributes thus to the safety of the cities down in the valley-bottom. Another problem is providing jobs, but here too plans are underway to arrange for commuting to Tarcento and expansion of the toy factory there in order to provide the jobs that are needed.

This letter to the editor is cited at such length because it expresses in the best imaginable way a whole series of quandries facing the reconstruction program for Friuli.

First let us take the more intelligible arguments.

(a) Italian decision-makers hesitate with reconstruction but begin to compete for the population's loyalty when the Protestant charity becomes active.

(b) The central government ('Rome') has too little detailed understanding of regionally important interactions between demographic ('depopulation') and ecological systems.

(c) The exploitation of marginal soils in the highlands is a voluntary and ecologically valuable service on the part of the mountain people in the interest of the general economy, arising from their devotion to their homes and deserving subsidies.

(d) Since highland agriculture, already uneconomic in itself, is even more irrational in the face of farm surpluses in the European Community, industrial jobs within a reasonable commuting distance must be provided.

Some of the points summarized here will crop up again in G. Valussi's formulation in Section 5.2. But let us first take note of a somewhat concealed argument in this very thought-provoking letter to the editor.

The people of South Tyrol, it is claimed, know best what mountain peasants need. 'The land handed down and tilled for centuries still means something there.' Our study also showed that ownership of home and land had a decisive influence on attitudes toward reconstruction. The question is whether this attitude can be mythologized the same way in Friuli as in South Tyrol.

The arguments of the donors of relief in Bozen were nourished by an undercurrent of associations deriving from another geographical situation.

In South Tyrol there was and is a conflict between the German-speaking natives and the Italians who migrated into the valleys (primarily Bozen and Meran) under Mussolini. Furthermore, the demographic task of compensating with high birth rates for the declining numbers of children in the fully urbanized South Tyrol population sometimes fell to the often prolific mountain peasant families ensconced in border areas on marginal soils. The mountain farms as 'fountains of youth' for the protection of majority status in communes that have become bilingual can, however, hardly be compared with the clusters of farms in the uplands of Friuli that are mostly inhabited by an over-aged remnant population. Nevertheless, arguments supposedly compatible with the donors' own perceptions have obviously played an important role as grounds for choice in finding where to locate relief activity. Citation of the altitudinal limit of settlements can be taken as an 'absolute value', even if justified today in ecological terms.

The thorough report of the Swiss 'National Volunteer Organization for Disaster Relief Abroad'[13] on their first reconnaissance mission to Friuli, submitted on May 21–25, 1976, gave the bases for later relief efforts in a much more reasoned manner but one resting ultimately on the same notions.

In view of the circumstances and special problems a reconstruction project was envisioned that in so far as possible would meet the following criteria.

(a) Reconstruction in a rather remote mountain zone hardly encompassed in the aid for reconstruction given at present or subsequently.

(b) Selection as to magnitude and locale of reconstruction of an objective within the technical, organizational, temporal, and financial competence of the volunteer service.

(c) Possibility of accomplishing a portion of the reconstruction before the onset of winter, so that if necessary the population can be safely carried through the winter in quarters temporarily somewhat restricted.

(d) Choice of a locality where the people are motivated to cooperate in reconstruction.

(e) Reconstruction layout corresponding sufficiently to established building and settlement patterns in the locality; where appropriate, Swiss building materials to be employed.

(f) Possibility of involving a large amount of local skill and labor force.

(g) Use of the opportunity provided by the reconstruction to give the local inhabitants encouragement and support in improving their economic situation in other respects as well.

A comparison among the eight communes and part-communes of Gemona, Osoppo, Buia, Oseacco, Prato, Subit, Cancellier and Attimis, on the basis of these seven principles resulted in a decision in favor of the Subit and Borgho Cancellier subdivisions of Attimis commune (CML U 59) (Fig. 4.10).

Subit and Borgho Cancellier are described thus.

The mountain village of Subit is at an altitude of 725 m about 45 km from

Figure 4.10 Switzerland concentrated its relief on a few small villages in the foothill area. It gave this aid not in the form of prefabs, but by constructing earthquake-proof permanent concrete buildings. For a group of 16 households, they endeavoured to find an architectural style derived from the house forms traditionally used in the mountains.

the Mediterranean coast at Trieste. It bears a great resemblance to our mountain villages in Tessino.

There are almost no more jobs in Subit village. The inhabitants in part have emigrated to find work abroad. One resident said that about 3 years ago 80% of the employable persons from Subit were working in Switzerland, mostly in the construction industry.

In the valley below, at an altitude of approximately 175 m, a large furniture factory was built that now employs many of the returned bricklayers or masons and some of the young people from the valley. Agriculture in Subit has been abandoned. Before the earthquake there were 3 cows in the village. People keep only small animals today.

The population is elderly. The mean age of the population is around 50.

It was the same for Borgho Cancellier.

The village is elderly (19 out of 33 inhabitants are over 60), probably because young people are working in Switzerland (15), or France and Belgium (13). These 28 emigrants will hardly return if they know that if they come home they will not have a roof over their heads.

The people interviewed wished to remain in the village.

The erection of new houses would make it possible for the young people to avoid moving to the valley where their jobs are (mainly the Patriarca furniture factory in Attimis).

The old part of the village will have to be given up entirely because it is in a landslide area.[4]

The comparison of geographical elements and components of the situation ('many similarities to our mountain villages in Tessino') known at home with those that show up again in the disaster area is characteristic of both this report of Swiss experts and the letter to the editor from the pastor in Bozen.

The question is whether superficial physiognomic similarities permit assumption of a structural equivalence. In neutral Switzerland, different standards of evaluation for what is agriculturally still a useful land may be applied because of the striving for autarky resulting from two world wars.[15] Farmland is highly prized there, with restrictions on its purchase by outsiders, and is in rather short supply because of the high proportion of unusable land in a high mountain country. Can standards worked out in a country with a GNP twice as high as Italy's (in 1972, $3940 *per capita vs.* $1960 *per capita*) be carried over as a guide for dealing with inhabitants of two subdistricts of Attimis? Or do these fellow mountaineers, the aid people from Switzerland and South Tyrol, understand the Friulian mentality better than the authorities of a central state seven times as big as Switzerland whose priorities do not include protection of an altitudinal frontier?

Thus the strong influence of perceptions and value standards upon relief measures from the outside is apparent: what is familiar to the members of a foreign relief organization as their own normative system and world view makes its own special demands on the relief effort[16] and outweighs the intentions the responsible local decision makers have for a particular commune or commune-fraction.

We have already tried to demonstrate, in Section 1.4, how such a 'distortion of the perception surface' can also arise for the incumbent authorities. Such distortions in the views of local and foreign aid-giving agencies suggested our taking a closer look at the problem of perception and assessment of natural hazards in a special study which follows in the next section.

4.4 Perceptions and assessments of natural hazards

Hazard research, as developed mostly in the United States in the mid-1950s, picks up the old theme of the interaction between society and nature at a point where the heightened dramatic quality of this interaction promises a deepened insight.

The traditional study of man's relationships in space concerned primarily *resources* – location and accessibility, soil and climate, energy supplies and natural resources evaluated and employed by man. Many works have treated the distinction between the ecumene and what lay beyond, gaining understanding of changes in spatial patterning of man–environment relationships from user-groups' discriminations between the useful and the no longer useful elements.

Hazard research also regards environment as resource but looks upon 'extremely great risk' as another characteristic over and above normal users' risks, such as game theory, for example, has been aware of for a long time. So hazard research begins when normal resources have failed us: all of a sudden in catastrophes, or slowly and insidiously in the case of gradual environmental degradation.

Hazards of this sort can take the form of floods or droughts, tornados or earthquakes, tsunamis or avalanches, landslides or volcanic eruptions. But such 'extreme events' become 'natural hazards' only when they represent not simply natural occurrences in half-empty, unused areas but exert their effects, usually unexpectedly, on individuals, societies and groups, disturbing and destroying their routines of living, so that people must find ways of 'adjusting' to them.

While it had been the purpose of our studies in the disaster area of Friuli to convey to the Italian authorities some conception of the need and feasibility of regional planning measures, hazard perception and adjustment, the assessment on the part of the local population could not be left in a 'black box'. We had to concern ourselves directly with the personal experiences of earthquake victims, and so we began by seeking out test-communes where we might expect to find such information in its most intensive form. This led us to two commune-fractions where the people could answer the following questions particularly well.

'How much do the people of the several districts struck by earthquake that were investigated know about the risks they live with?' 'How can the earthquake risk be distinguished from the rockfall and flood hazards that are always present?' 'How can such unquantifiable ideas as "home-loving" character or "rootedness" be brought to bear on ever present risks?' 'What part does the fear of a repetition of the catastrophe play in the life plans of the people?' 'Do they take a rational approach to the danger or is what happened accounted for in some irrational way?' 'How have the Friulians come to terms with the ever-present uncertainty of the earthquake?' 'What do the people concerned think of the authorities' efforts at reducing risks?'

To answer such questions, Michael Steuer worked as a member of the project on two particularly vulnerable commune fractions – Portis, north of Venzone, and Braulins to the west of Gemona (Fig. 4.11). These two communes are about 10 km apart as the crow flies, facing each other diagonally across the Tagliamento. Three different hazards jointly threaten them: flood, rockfall, and – as was dramatically demonstrated in 1976 – earthquake. This latter fact made it possible to compare two kinds of hazard, common everywhere in the mountains and already familiar to the people, with a new third kind. There have been such comparisons before. Golant and Burton[17] asked a group of 58 students to use a semantic differential in 1969 to compare 12 natural and manmade hazards. In 1971, Jackson and Mukerjee[18] asked 120 people in San Francisco about what adjustments it was possible to make to cope with earthquakes. But, as against such laboratory versions of hazard comparison, Friuli offered a real-life situation for investigation. Steuer talked with 140 heads of households in the two communes, representing about 60 per cent of resident families. He was interested not only in the actual behavior of the people of both places, but also in whether or not small geographical variations affected perception and behavior.

The two communes were very similar in their physical and human geography. *Portis* is located on the left bank of the Tagliamento between this dangerous stream itself and a steep, unstable cliff (Fig. 4.12). In 1977, it had 288 inhabitants in 99 households, 67 of whom replied to Steuer's questionnaire. The people's return to the former site was much debated for some time, until finally a different place was chosen for the village. Thus the population,

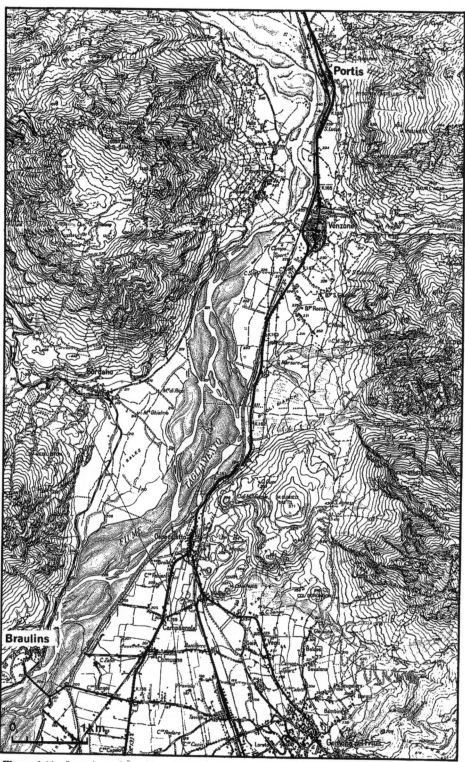

Figure 4.11 Location of Portis and Braulins in the Tagliamento Valley (courtesy of İstituto Geografica Militare, Carta d'Italia alla scala di 1:125 000: Moggio Udinese e Gemona del Friuli).

Figure 4.12 Broken loose by the earthquake of May 6 and numerous aftershocks, a hugh rock-fall decapitated the 'Spitz' above Venzone's district of Portis. It rolled over the church and grave-yard and blocked State Highway 13. Today, an artificial earth barrier defends this area, but the inhabitants do not trust it.

in the meantime living in a 'baraccopolis' beset with traffic noise, has to move a second time. *Braulins* lies 10 km further south on the right side of the Tagliamento on the steep eastward facing slope of Monte Brancot, whence descended a massive rockslide on to the village three days after the earth-quake, on May 9, 1976. At the time of the study, Braulins had 361 residents in 136 households, 75 of whom responded to Steuer's survey.

Braulins exemplifies clearly several elements of the situation. Retaining-fences 6 m high, equipped with elastic steel nets, (Fig. 4.13) attest to the fact that the rockfall risk has long been recognized in Braulins and that methods have been used to cope with it that worked well until the latest big earth-

Figure 4.13 Braulins, a district of Trasaghis, had experienced rockfalls before and therefore a huge fence was erected against them. It consisted of steel girders 6 m high, connected by an elastic steel net. The rockfall of May 9, 1976 tore this fence to pieces. Single boulders as large as 260 m³ (background) cascaded into the village. The wall that is to give shelter for the future is already overgrown by grass. The gravel shoals of the Tagliamento and the baraccopolis of Braulins can be seen in the background.

quake. Even so, an aftershock set off rockfalls that exceeded all previous experience. There were single blocks as big as 250 m³, weighing 500 tons. Their path from where they were detached down into the village was charted by Italian geologists. Down a 29° slope the blocks rolled as far as 400 m. In all, 25 000 m³ of material were displaced. The regional authorities therefore set up a heavy earth barrier against future rockfalls, that runs through the middle of the village and cuts off a section that is not supposed to be resettled from the part of the commune where people are allowed to rebuild, at a suitable distance of 27 m from this wall.

Whereas an unstable mountain cliff threatens Braulins on the west, the dikes carrying traffic to the narrow bridge to the east provide protection from potential floods on the Tagliamento, which overflowed its banks in 1963, 1966, and again in 1978. Torrential streams such as the Tagliamento are feared because of their highly erratic regimes.

Figure 4.14 summarizes the components of the situation once again: the unstable cliff, the catchment zone for rockfalls, the new earth barrier, the ruined village, the baraccopolis south of it, the Tagliamento, here spanned by

142

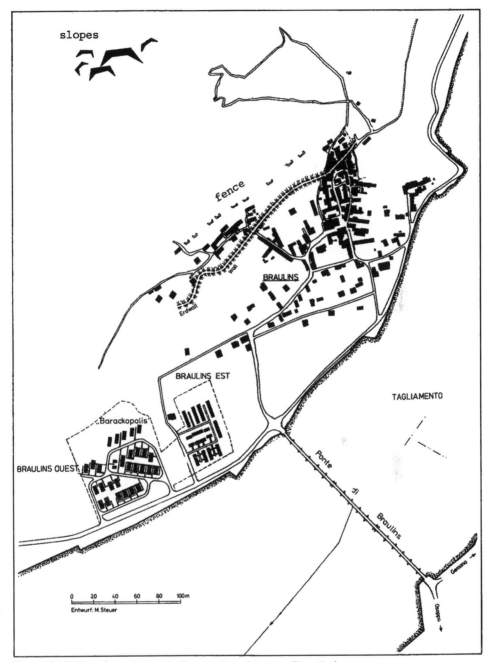

Figure 4.14 Location of the old village and prefab town (Braulins).

a small bridge where it narrows. Here, Steuer's questionnaire, administered by Friulian-speaking students from the Geography Department of the University in Udine (our partner institution), found a communicative populace.

Both communes investigated have elderly populations, and most of the employed persons work in the nearby industrial zone of Osoppo, and also run

Table 4.4 Social structure of the interviewed persons.

	Absolute frequency	Relative frequency
Occupation		
housewife	38	26.8
student	1	0.7
pensioner	50	35.2
unskilled worker	10	7.0
skilled worker	18	12.7
craftsmen, farmers	16	11.3
clerks, technicians	8	5.6
entrepreneurs, profess.	1	0.7
Education		
none	18	12.7
primary school	99	69.7
middle school	13	9.2
high school	10	7.0
university	2	1.4
Income (lire)		
less than 100 000	19	13.4
100 000–200 000	37	26.1
200 000–300 000	33	23.2
300 000–400 000	41	28.9
400 000–500 000	5	3.5
500 000–600 000	4	2.8

a small farmhold of their own on the side. Educational level and incomes are on a modest scale (Table 4.4).

The population of Braulins, with its rather peripheral location, is obviously more closely bound to its commune than that of the commune fraction of Portis, which sits on a main traffic artery (Table 4.5 & Fig. 4.15). The populations of both communes are highly conscious of the natural hazards there, with rockfalls and floods a little more clearly perceived in Braulins than in Portis (Table 4.6).

The causes of different degrees of damage are variously evaluated (Table 4.7). Thus, the faulty method of construction is more blamed for damage in Portis than in Braulins. Correspondingly, earthquake-proof build-

Table 4.5 Comparison of selected socio-demographic responses of people in Portis and Braulins.

	Portis (%)	Braulins (%)
lifelong residence in settlement	62.7	81.3
over 30 years' residence	62.7	77.3
no intention of moving	80.6	85.3
no desire to live elsewhere	53.7	77.3
satisfied with job conditions	29.9	44.0
living conditions rated satisfactory to good	47.8	66.6
moderate-to-strong home links	67.1	89.4
moderately-to-well integrated	49.2	73.4

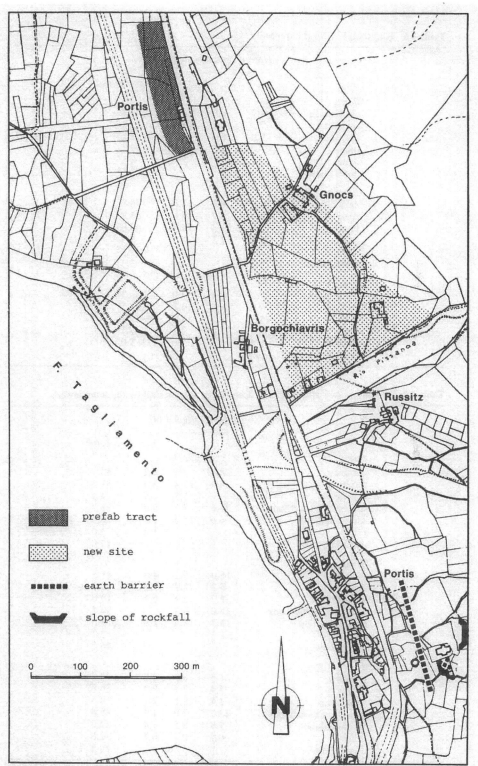

Figure 4.15 Old Portis between floods and mountain rockfalls; and prefab tract between the railroad tracks and highway. Planned new site.

Table 4.6 Knowledge of local hazards.

	Hazard rank assigned			Total mentioned
	1	2	3	
Portis				
earthquake	64.2	9.0	1.5	82.1
rockfall	7.5	26.9	13.4	55.2
flood	9.0	26.9	13.4	55.2
other		4.5	4.5	13.4
none				4.5
don't know				7.5
Braulins				
earthquake	52.0	8.0	12.0	78.7
rockfall	12.0	26.7	16.0	61.3
flood	13.3	33.3	14.7	68.0
other	1.3	1.3		2.7
none				9.3
don't know				5.3
Total				
earthquake	57.7	8.5	7.0	80.3
rockfall	9.9	26.8	14.8	58.5
flood	11.3	30.3	14.1	62.0
other	0.7	2.8	2.1	7.7
none				7.0
don't know				6.3

Table 4.7 Specific causes of earthquake destruction as perceived by the interviewees.

	Rank assigned (4)			Total
	1	2	3	
Portis				
intensity	28.4	7.5	6.0	53.7
duration	4.5	13.4	3.0	31.3
repetitions		1.5	1.5	4.5
construction methods	31.3	7.5	9.0	53.7
site foundation	6.0	11.9	1.5	20.3
other			1.5	3.0
don't know				17.9
Braulins				
intensity	16.0	5.3	4.0	41.3
duration	5.3	2.7	1.3	22.7
repetitions	2.7	2.7	2.7	12.0
construction methods	16.0	6.7		38.7
site foundation	13.3	5.3	1.3	24.0
other	1.3			2.7
don't know				26.7
Total				
intensity	21.8	6.3	4.9	47.2
duration	4.9	7.7	0.7	26.8
repetitions	1.4	2.1	2.1	8.5
construction methods	23.2	7.0	4.2	45.8
site foundation	9.9	8.5	1.4	22.5
other	0.7	0.7	0.7	2.8
don't know				22.5

ing techniques are given more weight as conceivable adjustments there. The wall is more talked about in Portis. But is is evidently not trusted, as the removal of the Portis community from the potentially protected area behind the wall to another part of the commune reveals (cf. Tables 4.8 & 9).

People's expectations vary concerning repetition of the various sorts of catastrophes, but everyone agrees that another rockfall will be the first type to occur (Table 4.9). The reason for this could be that the Italian government has been trying to do something about this danger by building the earth barrier, and so has evidently increased the people's awareness of the risk. Table 4.8 shows the knowledge of adjustments in such a situation.

The residents' thoughts about the catastrophes were tested with four pairs of statements contrasting fatalism with the attitude that a person is completely responsible for his own behavior. The values obtained for both communes showed no significant differences between them and a large preponderance in favor of external control. So some of the statements in Table 4.8 about adjustments must be seen in the light of the fact that the overall opinion tends to transfer responsibility to a higher power (Table 4.10).

Italian research in geotectonics and seismology has lately been aimed at finding a more precise way of calculating recurrence probabilities on the basis of investigations such as those being conducted especially in the Osservatorio Geofisico Sperimentale at the Grotto Gigante site in the Karst near Trieste. They show the Portis fraction of Venzone to lie in the most endangered zone; the Braulins fraction of Trasaghis, in an immediately adjacent area, must reckon with a repetition of an earthquake of intensity IX within from 60 to 100 years (Fig. 4.16).

We may relate these measurement data to the population's expectations and fears. The comparison of statistically established intervals of recurrence over a documented earthquake history extending back more than 1000 years in Friuli reveals four expectation patterns with regard to recurrence (Table 4.11). The large number of 'optimists' who underestimate the risk of a recurrence undoubtedly represents a major portion of the population, officials included. One indicator is the construction but then blowing up of the brand new ten-storey Central Hospital in Gemona in June 1979, that had been so badly damaged before its completion on May 6, 1976 that it was a safety hazard. The graph in Fig. 4.17 displays what the people know about occurrences and causes of the three hazards investigated, and how they estimate recurrence intervals. Differences showed up here between the populations surveyed in Portis and Braulins with regard to perceptions and attitudes, experience and adjustment capacity, knowledge of protective measures, expectations of recurrence, and a series of other factors. Although closely similar socially and demographically, the more isolated population of Braulins, as a rule, proved to be better integrated into their surroundings, accepting of their fate (cf. external controls) and less well informed than Portis people with their wider contacts. Thus, even small locational differences can evidently affect the cognitive and affective attitudes of a given population and evoke a different behavior. In Fig. 4.18, the residents of Portis are always a little bit 'better off' in their scores for knowledge of hazards, adjustments and overall rationality.

Whereas it is this sort of attitude measurements that are most important for reconstruction planning, hazard research is likewise interested in the charac-

Table 4.8 Knowledge of adjustments.

Adjustments	Portis				Braulins				Total			
	Rank assigned (%)			Total %	Rank assigned (%)			Total %	Rank assigned (%)			Total %
	1	2	3		1	2	3		1	2	3	
Private												
quake-proofing	61.2	3.0	1.5	70.1	44.0	12.0		60.0	52.1	7.7	0.7	64.8
insurance	1.5	4.5	1.5	9.0	2.7	1.3		5.3	2.1	2.8	0.7	7.0
live elsewhere in commune	1.64	17.9		38.8	18.7	6.7		29.3	17.6	12.0		33.8
leave commune for good	7.5	3.0		10.5	10.7			12.0	9.2	1.4		11.3
other private	1.5			1.5	1.3	2.7		4.0	1.4	1.4		2.8
Public												
build wall	32.8	4.5	1.5	50.7	22.7	5.3		37.3	27.5	4.9	0.7	43.7
catchment fences	4.5	3.0		11.9	6.7	1.3		9.3	5.6	2.1	0.7	10.6
blow up cliff	9.0	6.0	3.0	23.9	9.3	1.1	1.3	13.3	9.2	3.5	2.1	18.3
build dikes	6.0	14.9	4.5	37.3	8.0	9.3	2.7	29.3	7.0	12.0	3.5	33.1
zone for risk	11.9			14.9	2.7	1.3		8.0	7.0	0.7		11.3
government aid	3.0	6.0		10.4	2.7	2.7	2.7	10.7	2.8	1.4	4.2	10.6
other public	1.5	1.5	1.5	7.5	4.0	2.7		13.3	4.2	2.8	2.1	10.6
none				4.5				2.7				3.5

Table 4.9 Expectations concerning repetition of the various hazards by per cent of responses.

| | Period within which repetition expected (years) | | | | | | | | | |
| | <10 | | | | | >100 | | | | |
	5	5–10	10–20	20–50	50–100	100–200	200–300	>300	Never again	Don't know
Portis										
earthquake	46.2	11.9	0	1.5	10.4	9.0	7.5	10.4	11.9	37.3
rockfall		5.9	2.9				5.9		7.4	29.9
flood	26.8	5.9	7.4	4.4	2.9		3.0		7.4	44.8
Braulins										
earthquake	48.0	10.7	2.7	6.7	14.7	4.0	9.3	9.3	22.7	20.0
rockfall		6.6	2.6	1.3			5.3			20.0
flood	30.6	10.6	10.6	1.3	2.6		1.3	14.6	18.6	29.3
Total										
earthquake	47.2	11.3	1.4	4.2	12.7	6.3	8.5	9.9	17.6	28.2
rockfall		7.0	2.8	0.7	1.4		5.6		10.6	24.6
flood	28.9	8.5	7.7	2.8	1.4		0.7		13.4	36.6

Table 4.10 External and internal control.

	Braulins	Portis
external control	50.7%	48.5%
undecided	25.3%	26.9%
internal control	24.0%	23.6%

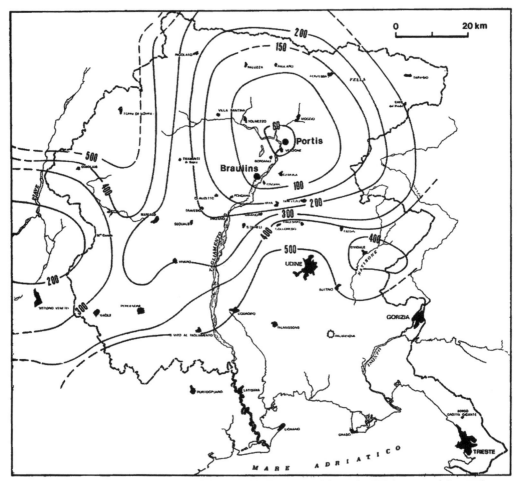

Figure 4.16 Recurrence probabilities for earthquakes of intensity IX in years (observation period = 150 years) (source: Osservatorio Geofisico Sperimentale Trieste 1978).

Table 4.11 Expectation patterns with regard to recurrence.

	Percentage of people expecting recurrence within			
	50 years (pessimist overestimates)	50–100 years (realist)	more than 100 years (optimist underestimates)	'Don't know' (uncertain)
Portis	13.4	10.4	38.8	37.3
Braulins	20.1	14.7	45.3	20.0
average	16.9	12.7	42.3	28.2

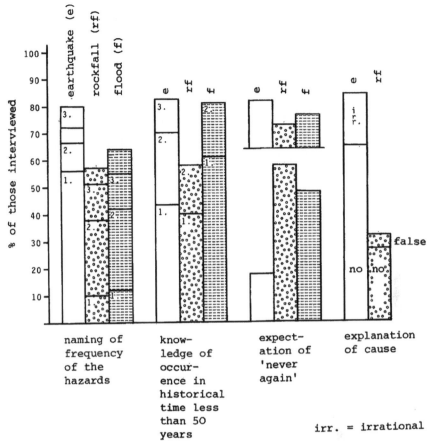

Figure 4.17 Comparison of knowledge about frequency of occurrence, causes and possible recurrence of three hazards (1, 2, 3 = ranks assigned).

teristic features of various hazards perceived by a population in ways evoking varied semantic images. Steuer was able in this instance to extend the findings that Golant and Burton had arrived at in laboratory experimentation.

Particular hazards almost always have a similar profile in both communes (Fig. 4.20). Comparing them, the earthquake profile leans more strongly toward the irrational, being considered inscrutable and irregular, but, on the other hand, there is the hope it will not happen soon again.

In the same way that each type of event possesses its own image, the different sorts of adjustments to them held possible are variously conceived. The observations of American investigators are to the effect that technological, capital-intensive adjustments undertaken by public agencies make more sense to people than do 'harmonious adjustments' such as changes in land use and safety margins against danger sources, or measures that depend on personal decision and individual effort, such as taking out insurance.

Within both commune fractions, zones of higher risk from both rockfalls and floods could be areally delimited and the statements of their former residents compared. Each such interview was keyed to residence site in the

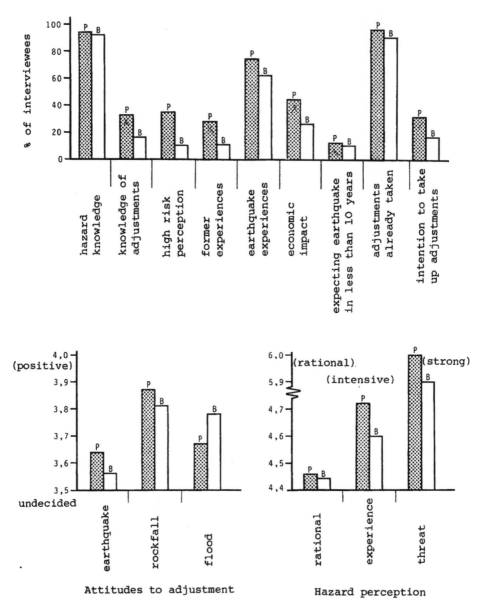

Figure 4.18 Differences between the populations surveyed: P = Portis; B = Braulins.

ruined village. Having people who were now living in the barracks settlements think back to where their former homes (sometimes to be reconstructed) had been, was intended to show whether their perceptual space covered the whole commune fraction as one unit, or instead reflected in precise detail the exact location of their own property. Then the level of perceived threat was noted, and related to the distance of the home from the threat source, either the river or the cliffs. It was found that the people who had formerly lived further away from the slopes and river banks thought little differently from those directly endangered. So, whereas locational differences between commune fractions

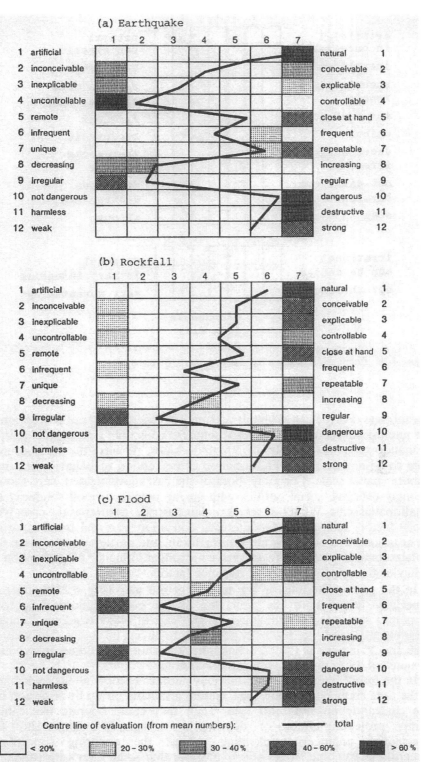

Figure 4.19 Evaluation of three different hazards: (a) earthquake; (b) rockfall; (c) flood.

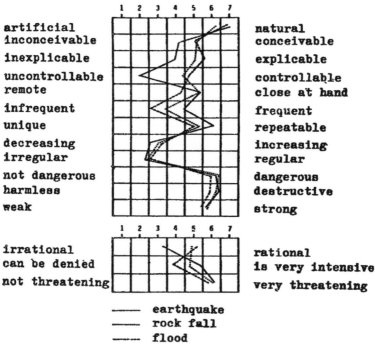

	1	2	3	4	5	6	7	
artificial inconceivable								natural conceivable
inexplicable								explicable
uncontrollable remote								controllable close at hand
infrequent								frequent
unique								repeatable
decreasing irregular								increasing regular
not dangerous harmless								dangerous destructive
weak								strong
irrational can be denied								rational is very intensive
not threatening								very threatening

------ earthquake
·········· rock fall
——·—— flood

Figure 4.20 Comparison of three hazards in Portis and Braulins.

(Braulins *vs.* Portis) can probably account for dissimilarities of perception, on the micro-level the whole commune seems to constitute a perceptual unit, but a smaller spatial area of 100 or 200 m does not. At least, the methods available could not detect a differentiation of perception or behavior at a more minute spatial scale. Certainly, both of the cases investigated were those of strongly cohesive social entities with strong neighbourhood solidarity at a small spatial scale. What this seems to demonstrate, as often is the case, is that social factors such as similar interests, education, external controls, attachment to local symbols, degree of integration, and participation in local face-to-face communication networks have a stronger influence on perception and behavior than microgeographic distance factors.

In the part of Braulins in the middle of the map (Fig. 4.21b), near and especially just behind the wall, the strong concentration of interview responses that express willingness on the part of many residents to change their place of residence probably goes back to having no choice. The wall, 7 m high and 20 m wide at its base, constitutes the limit for building, from which a distance of 27 m on the town side is required.

In the future, the protective walls may perhaps become a manmade hazard if the area contained behind them should gradually fill up by accretion from the constantly unstable cliff face. Then they could operate like catapult ramps, projecting chunks of rock into the village. This ranges short-term speculative engineering planning against the enduring concerns of generations to come. It is entirely possible that some later catastrophe will wreak more havoc because so much more has been invested in building as a

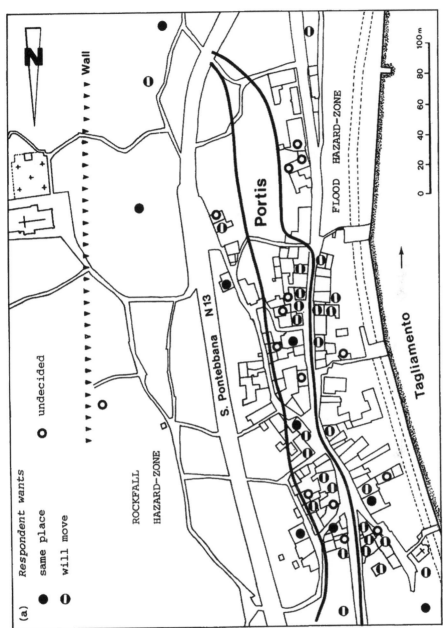

Figure 4.21 Spatial distribution of willingness to move: (a) Portis.

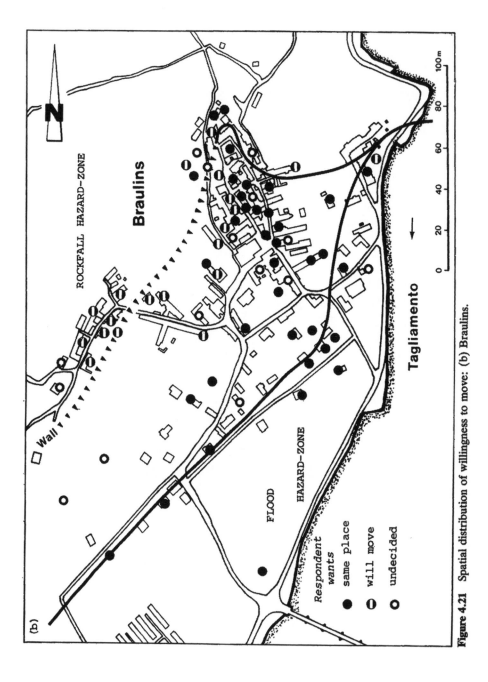

Figure 4.21 Spatial distribution of willingness to move: (b) Braulins.

Figure 4.22 Rockfall and flood areas in Friuli (source: *Piano urbanistico* 1976, Fig. 10).

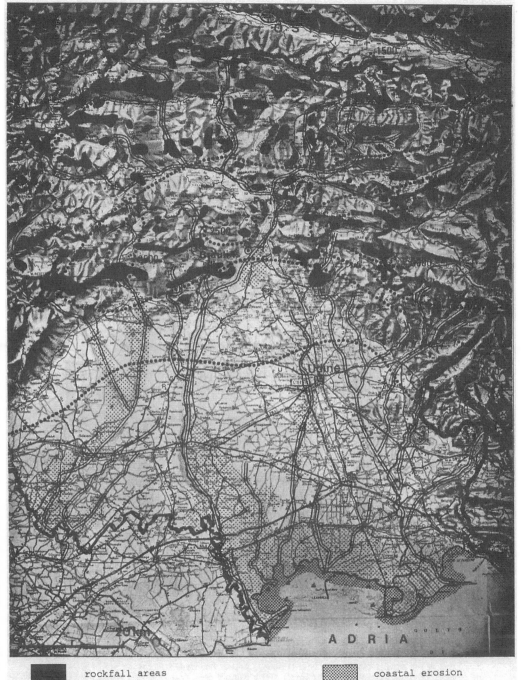

▮	rockfall areas	▨	coastal erosion
▨	flood areas	••••••	isohyets 1921–50

Table 4.12 Indicators of attachment to commune.

| | Friulians | Italians | Europeans | None | Other | 'At home' in commune | |
						Yes	No
Portis	65.7	11.9	10.4	9.0	3.0	89.6	10.4
Braulins	73.3	6.7	5.3	10.7	2.7	94.7	4.0
total	69.7	9.2	7.7	9.9	2.8	92.3	7.0

result of the supposed protection afforded by the wall than would have been without it. Even apart from latent earthquake risk, the narrow valleys of Friuli will continue to be risky areas for settlement because of high relative relief and copious precipitation (Fig. 4.22). Why do people cling to areas recognized as so dangerous anyway? Philip W. Porter cites some reasons:[19]

(a) they have no choice, other people will not permit them to live elsewhere;
(b) there are advantages or opportunities in the area which are sufficiently important to them that they are willing to assume the risk of hazards also present there;
(c) and they were born there or for other reasons have formed strong attachments to their locality, and prefer their place to any other.

Such an emotional attachment to place is especially well developed in Friuli. Asked for their national identity, the respondents designated their ethnic groups as shown in Table 4.12.

Short-range migration certainly takes people out of range of rockfall or flood danger zones, but it cannot prevent the spatially extended threat of

Figure 4.23 The (Italian) inscription 'Portis shall be reconstructed here' was put up after the first quake. Later rockfalls chipped away the 'QUI' ('here') and blocked the road. On the left is one of the smaller blocks. The later (Friulian) poster 'Puàrtis al scûen torna SU' repeats once more this defiant desire of 'SU' ('here'). But it will not come true. The new Portis will be built elsewhere (cf. Fig. 4.15).

possible further earthquakes. Moving further away, however, would compel Friulians to give up their language and cultural cohesiveness, the sense of unity. Fear of assimilation and Italianization evidently figures as an overriding hazard (Fig. 4.23).

In summary, Steuer's study[20] is an attempt to compare three different hazards and adjustments considered appropriate, in a real life situation fresh in memory, in two similarly exposed commune fractions, and to relate causal elements to them. In supplementary studies it is often shown that the tensions between fears and hopes revealed in this study also strongly influence the mobility behavior of the people and the investment activity of entrepreneurs, while particular emphasis should also be given to the strong tendency of the population to identify with their ancestral homeland, which provides their determination to remain.

Notes

1 Barbina, G. 1976. Il Friuli centrale dopo gli eventi sismici del 1976. *Boll. Soc. Geog. Ital.* **Series X, VI** (10–12), 607–36.

2 Glauser, E., H. Gugerli, E. Heimgartner, B. Rast and B. Sägesser. 1976. Das Erdbeben im Friaul vom 6 Mai 1976. Beanspruchung und Beschädigung von Bauwerken. *Schweizer. Bauzeit.* **94** (38).
Heimgartner, E. and E. Glauser, 1977. Die Erdbeben im Friaul zwischen dem 6 Mai und dem 15 September 1976. *Schweizer. Bauzeit.* **95** (12).

3 The newly founded journal *'Ricostruire'* (*'Reconstruction'*) dedicated its issue 1 of April 1977 to a documentation of the building program for Friuli. Table 4.1 is taken from the paper of L. Di Sopra: La baraccopoli più grande d'Europa, documento sulla tipologia delle baracche in Friuli, p. 5.

4 Mitchell, W. A. 1976. Reconstruction after disaster: The Gediz earthquake of 1970. *Geog. Rev.*, 296–313.

5 Mitchell, W. A. op. cit., p. 311.

6 Mitchell, W. A. op. cit., p. 313.

7 Pagnini, A. M. P. 1975. *La zona terremotata di Dasht-e-Bayaz* (*Iran Orientale*). University of Trieste, Facolta di Szienze Politiche, Publication no. 4.
Centro Studi e Iniziative 1968. *Piano di sviluppo democratico per le valli Belice, Carboi, Jato*. Particinio.

8 Meneghel, G. 1976. *Indagine socio-geografica sui movimenti migratori nel comune di Lusevera* (*Provincia di Udine*). Contributi geografici allo studio dei fenomeni migratori in Italia. Facolta di Lingue, Udine.

9 Deutsches Rotes Kreuz, Schweizerisches Rotes Kreuz, Caritas Schweiz, Caritas Kärnten, Deutscher Caritasverband, Diakonisches Werk für Österreich, Hilfswerke der Evangelischen Kirchen der Schweiz, Arbeiterwohlfahrt Bayern, Schweizer Arbeiterhilfswerk, Staatliches Schweiterisches Freiwilligenkorps für Katastrophenhilfe im Ausland und Diakonisches Werk der evangelisch-lutherischen Kirche Bayerns (Red Cross organizations of Germany, Switzerland and Austria, Caritas organizations (catholic) and Protestant relief organizations, labor aid associations etc. of the three German-speaking neighbor countries).

10 Lindenmeyer, H. O. G., Vicar of Bozen, 1977. Die Bergbauern von Cesariis. *Münchner Gemeindeblatt. Sonntagsblatt für die Evang.-lutherische Kirche in Bayern* no. 15, 10 April, p. 18.

11 Holzhaider, H. 1977. Das alte Friaul wird es nicht mehr geben. *Münchner Gemeindeblatt.* No. 11. p. 8.

12 German-speaking government of the Autonomous *Province* of South Tyrol (capital Bozen/Bolzano) which in turn is part of the Autonomous *Region* of Trento–Alto Adige (capital Trento), which can be compared to the Region of Friuli–Venezia–Giulia.

13 I am warmly appreciative of the support given by P. Studer, delegate of the Swiss Bundesrat for the 'National Volunteer Organization for Disaster Relief Abroad'.

14 Manuscript, pp. 15–19, Bern 1976.

15 Sehmer, I. 1959. Studien über die Differenzierung der Agrarlandschaft im Hochgebirge im Bereich dreier Staaten. *Münchener Geographische Hefte* **17**. Regensburg: Kallmünz.

16 The amount of all foreign aid given is estimated by R. Strassoldo at a total of $95 million (see Strassoldo, R. and B. Cattarinussi (eds) 1978. *Friuli: la prova del terremoto*, 196–7. Milan).

17 Golant, S. and I. Burton 1979. *The meaning of a hazard. Application of the semantic differential*. Natural Hazard Res. Working Paper 7, Toronto.

18 Jackson, E. L. and T. Mukerjee 1974. Human adjustment to the earthquake hazard of San Francisco, California. In *Natural hazards — local, national, global*, G. F. White (ed.). New York: Oxford University Press.

19 Porter, P. W. 1978. *The ins and outs of environmental hazards*. Working Paper 3, University of Minnesota.

20 Steuer, M. 1979. Wahrnehmung und Bewertung von Naturrisiken am Beispiel zweier ausgewählter Gemeindefraktionen im Friaul. *Münchener Geographische Hefte* **43**. Regensburg: Kallmünz.

5 Conclusions

5.1 Assessment of the catastrophe by local planners

In the foregoing chapter, we attempted to show how the population of two small, especially hard-hit commune fractions perceived the hazards threatening them, how they reacted to them, and what they thought of government protective measures. They are all amateurs in this matter of risk estimation and can only draw on popular beliefs for assessments of the future and of reconstruction of their homeland.

Contrasting with them are the professionals, the planners in charge of reconstruction. They were only recently recognized as key figures for the future shaping of a region by Ciborowski (1967)[1] and Douard (1978)[2]. What they provide is mainly testimony on the consequences of a catastrophe, particularly an evaluation of disruption and gain in the sense of Kates' formula (1976)[3] (Fig. 5.1).

In assessing disruption and gain, two hypotheses can be set up:

(a) catastrophes heighten regional and intergroup inequality;
(b) reconstruction is less influenced by extent of destruction and more by

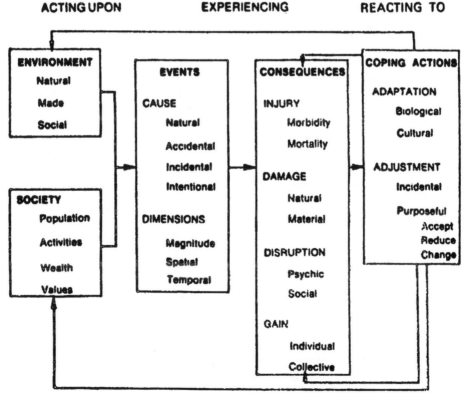

Figure 5.1 The environment as hazard (source: Kates[3]).

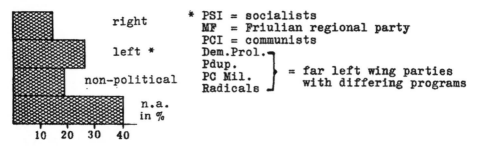

Figure 5.2 Political inclination of planners interviewed.

factors such as development trends, regeneration potential, plans, laws, subsidies, personal expertise (especially of planners) and the attitudes of the people affected. (Stagl 1980, p. 10).

In order to test both these hypotheses, Rudolf Stagl[4] interviewed 95 planners in the 45 planning offices of communes in the 'destroyed' category – a group that had just emerged as a profession because of the catastrophe, and was correspondingly youthful: 40 per cent of those interviewed were under 25, almost 80 per cent under 30; most had graduated from an institution of higher learning, themselves lived in the destroyed communes, and had qualified for planning work in the emergency situation.

In Sections 4.2 and 4.3 we have already shown how strongly political concepts affect reconstruction planning. So it is important to know how the planners interviewed place themselves politically – apart from the 40 per cent who declined to answer such a ticklish question (Fig. 5.2). The better educated much more frequently reported themselves as left-wing and were more willing to state their political preferences openly. Left-wing persons tended to exaggerate problems, right-wingers to minimize them.

In this inquiry a ranked list of problems of the communes was created, falling neatly into three groups according to their varying urgencies (Table 5.1). The pressure of time, bureaucratic problems and the need for field consolidation ranked highest. Field consolidation is presupposed by any reconstruction, but encounters the problems of properties in small parcels in

Table 5.1 List of problems (1 = severe difficulties; 2 = slight; 3 = no difficulties).

time pressure and public impatience	1.26
bureaucracy in lower and higher levels	1.27
field consolidation	1.28
co-operation with citizens	1.67
lack of skilled workers in commune	1.68
lack of juridical basis for planning measures	1.74
apportioning of national and regional subsidies	1.79
conflicts of competences	1.84
no working facilities for the planners	1.84
coordination with other agencies/collaboration	1.86
emigration – people abroad	1.87
external efforts to influence planning	1.92
lack of specific ideas about future commune	2.01
political strife within commune	2.12

Friuli and the high prestige attaching to land ownership. Reports of similar hurdles from communist Yugoslavia attest that these are not just capitalistic factors that interfere with planning.[5]

> The disaster area of Montenegro: bureaucratic hurdles and the mentality of the people meant that reconstruction only proceeded with faltering steps . . . Time was richly squandered too because almost nothing happened in the first months . . . This caution of the officials cripples any sort of private initiative. Even people with money and able to work can neither build nor repair without a permit from the commune.

Such evidence led Stagl to propose that communes in districts threatened by catastrophes should try to assemble larger land parcels in the possession of public agencies by trading or purchasing, which in normal times can serve as parks, playgrounds, or parking lots, and can be ready immediately after a catastrophe to be used as reception areas for building emergency housing. For often the loss of a parcel by requisitioning for a baraccopolis is regarded by its owners as worse than losing their home in the earthquake.[6]

This almost superstitious clinging to the land and the method of apportioning indemnities favoring individual applicants leads to a proliferation of single-family residences and a decrease in the number of multiple units with rental space, thus putting pressure on the land market, reducing the density of the built-up area, and causing towns to spread out further and settlement to scatter on to previously open lands. Planners look askance at this, because it greatly raises the cost of communal services (streets, water supply etc.), distances increase, and the characteristic closed character of Friulian towns and villages gives way to a mushy settlement form: 'Friuli is going to look like another suburb of Milan.'

Since purposely pessimistic estimates inflated the damage level, large amounts of public funds flowed into the disaster area which, particularly in the middle of the destruction zone (Fig. 5.3), produced satisfaction among the planners, while peripheral districts were complaining about being neglected.

An estimation of the financial resources is shown in Figure 5.4.

Just as there is regional preference and discrimination, so reconstruction also sharpens social inequalities among various groups (Fig. 5.5).

According to the planners, who is it who most *profits* from a catastrophe?

(a) Industry, trade and handicraft, by way of expansion credit and increased orders;
(b) former renters, who get their own homes;
(c) families with several wage earners, and plenty of determination, who began rebuilding promptly, received higher subsidies, and lost less to inflation;
(d) a minority of owners of several houses who modernized, then rented them.

The reconstruction process was *disadvantageous* to:

(a) old people, financially weak, with little initiative and no connections;
(b) a minority of owners of several houses whose renters became independent;

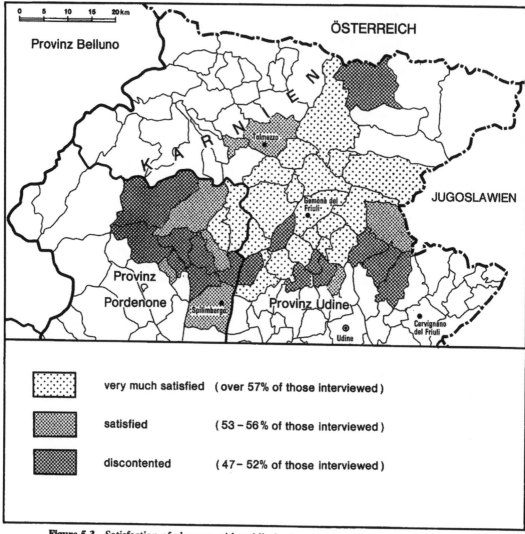

Figure 5.3 Satisfaction of planners with public funds.

(c) people who did not come back at the right time and who 'fell through the mesh' of the laws.

These observations can be represented in a diagram (Fig. 5.6).

Stagl made a division based on location, degree of damage, and size of commune. Since various groups in these communes blame the same regional and national laws for their being either better or worse off, an accentuated polarization with regard to loyalty versus alienation from government among the citizens becomes evident. A person who wants the government to take charge of everything, such as the planners reporting themselves as 'left wing', blames the advantages and disadvantages for various groups that have become apparent on the laws, and thus on government policy. On the other

Figure 5.4   Estimation of financial resources.

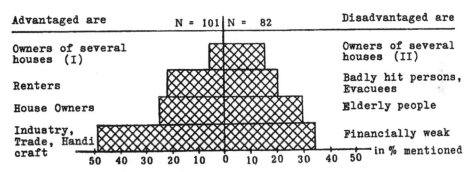

Figure 5.5   Groups advantaged or disadvantaged.

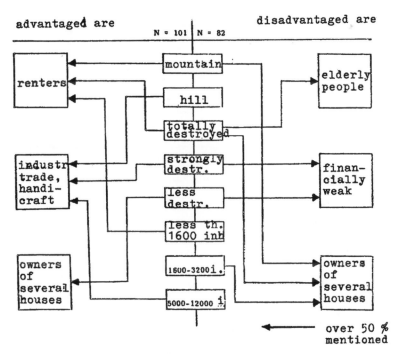

Figure 5.6   Advantages and disadvantages in regional view.

hand, someone who believes that an ambitious individual will make his own way through private initiative sees less of advantages and disadvantages in government policy.

Meanwhile, there is an interesting regional difference in satisfaction with both Law 30, concerning allocations of subsidies for repairs, and Law 63, governing the complete rebuilding of homes (Fig. 5.7). They were enacted on June 20 and December 23, 1977, respectively. Law 30 promotes private repairs up to 80 per cent with no repayment required, and new buildings can be subsidized up to 100 per cent. This obviously works better in the less-harmed south of the disaster area than in the hard-hit north. Kates is right, regionally speaking, when he says 'Disaster is not a great leveller . . . and socio-economic groups will be determined after disaster as they were before, with the rich still rich and the poor still poor'.[7]

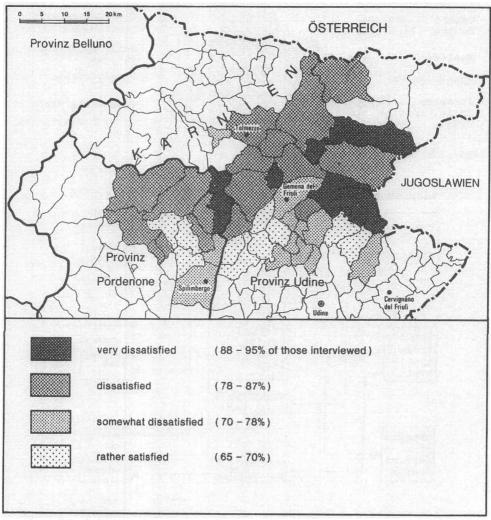

Figure 5.7   Opinions on reconstruction Laws nos. 30 and 63.

166

The time factor is especially important here. Whoever got started sooner could build more cheaply. Since the rate of price increases from October 1, 1977 to June 1, 1978 reached 15 per cent in the construction sector, the regional administration had to adjust subsidies to the inflation rate, and on November 1, 1978 raised them by about 18 per cent. But meanwhile inflation had climbed yet further. So early starters can build themselves bigger houses than the latecomers, and older and poorer people watch social differences get greater and greater with every year of reconstruction, as they stare out of their barrack windows.

To be sure, this difference in the promptness of reconstruction and the greater satisfaction with the reconstruction laws in the southern part of the disaster area could naturally reflect some psychological block: anyone undertaking reconstruction does so wondering, 'Does it all make sense? What happens after the next earthquake?'

When a person takes the danger of earthquakes more seriously – which happens more in the north than in the south – he is more aware of his responsibilities in planning and rebuilding, safety measures are more scrupulously observed, setbacks follow the bylaw procedures, and people take longer to decide on the final reconstruction plan. But in time, a process of repression sets in,[8] in the north as well. At this point the estimation of the danger of repetition of earthquakes becomes differentiated significantly according to the size and position of the commune. In larger communes in the hill country, the risk is much more strongly underestimated than in small mountain communes.

The map in Figure 5.8 shows estimates of earthquake danger made by the 95 planners. Most of them believe that the quakes of 1976 have been a singular event in their lives and will never be repeated. Concerning the likelihood of a strong new quake, only two-thirds expressed an opinion (Fig. 5.9). This appears to depend on present extent of destruction in the commune, opportunities for aid and development (i.e. size), and position (with respect to major thoroughfares).

Arranging communes into contrasting pairs, an earthquake of the severity of that of 1976 is expected in mountain communes considerably sooner (in 62 years) than in hill communes (100 years); in medium-sized and large communes with over 1600 inhabitants, very much later (in 99 years) than in small communities (60 years); in very badly destroyed communes, sooner (in 65 years) than in badly and less badly destroyed ones (97 years) (Fig. 5.10). Thus, beside the actual extent of damage, other characteristics of a commune that have nothing to do with the earthquake itself also play their part in the assessment of likelihood of a repetition – including attitudes such as optimism and pessimism on the part of planners and the population that are hard to deal with. Thus we can also regard the variable 'supposed likelihood of the recurrence of an earthquake' as an indicator for the strength of feelings about losses suffered. It goes a long way toward justifying, at a late stage, the designations of extent of damage suffered that we have already discussed (cf. p. 38). At the same time, the estimate of the still existing danger becomes a measure of the degree of success in reconstruction. In this regard, the small mountain communes show up as having been most crippled by the aftereffects of the earthquakes.

The planners were questioned about their own expectations concerning a

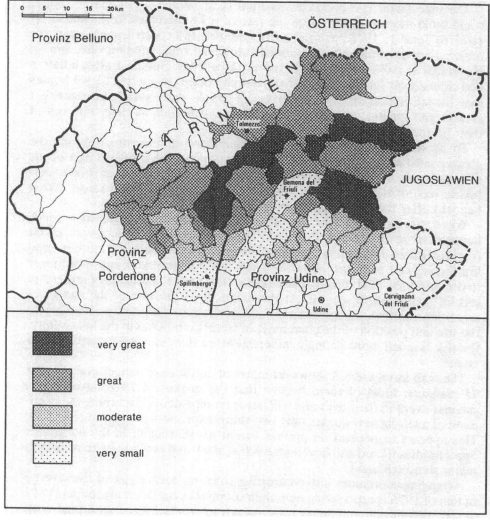

Figure 5.8 Estimations of earthquake risk.

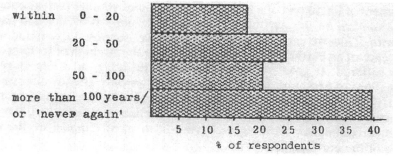

Figure 5.9 Opinions concerning the likelihood of a strong new quake.

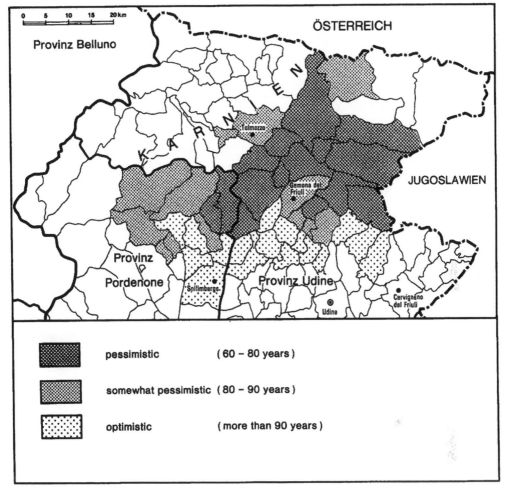

Figure 5.10 Expectation of frequency of recurrence of strong quakes.

recurrence and about how well the population of their present communes could estimate the risk. Comparison of the studies of Stagl and Steuer[9] allows us to assess three different opinions. This comparison shows that the planners, themselves all too optimistic, nonetheless still underestimate the optimism of the population.

When Ambrasey's scale[10] is compared with the events that took place *after* 1976, the 'series rule' is refuted (Table 5.2). That is to say, under that rule the earthquake of September 15, 1976 should have occurred not after 4½ months but 76 years later. The earthquake of September 17, 1977, at 6.0 R, should have happened only 56 years later. The earthquake of April 18, 1979, at

Table 5.2 Recurrence of earthquakes in Friuli.

strength (Richter scale)	6.75	6.50	6.25	6.00	5.75	5.50	5.25	5.00	4.50	4.00
recurrence (years)	587	253	116	56	29	16	9	6	3	1

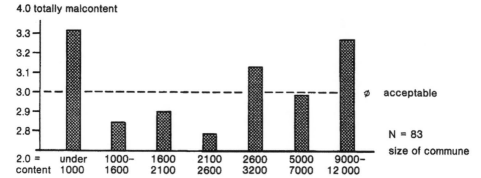

Figure 5.11 Community size and citizen's satisfaction with administration.

4.8 R, should have happened 5 years later. Since the last three mentioned were regarded as aftershocks and looked on as a unit, the population is reposing in a false sense of security, looking upon precautionary measures to an increasing degree as superfluous and on stricter building regulations as a bureaucratic annoyance.

This dissatisfaction with a too meticulous commune administration expresses itself especially in the small, and therefore helpless, mountain communes and in the large, sprawling and inherently heterogeneous communes such as Tarcento and Gemona (Fig. 5.11).

The reasons for dissatisfaction are mainly connected with the points shown in Figure 5.12. Because commune planners cannot be held equally responsible for all these causes, many of them feel they are scapegoats. Also, the planners criticize the abuse, common among the affected people, of asking for more than they need in order to receive enough.[11] Again here, regional differentiation of the dissatisfaction of commune residents with the planners is evident, calling attention to structural differences in the nature of the problems (Fig. 5.13). However, the citizen is dissatisfied not merely with his own communes, but also with the regional government and its policies.

The reasons for this dissatisfaction are mainly connected with the following points:

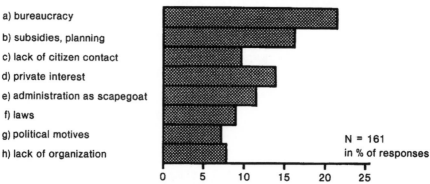

a) bureaucracy
b) subsidies, planning
c) lack of citizen contact
d) private interest
e) administration as scapegoat
f) laws
g) political motives
h) lack of organization

N = 161
in % of responses

Figure 5.12 Reasons for dissatisfaction.

170

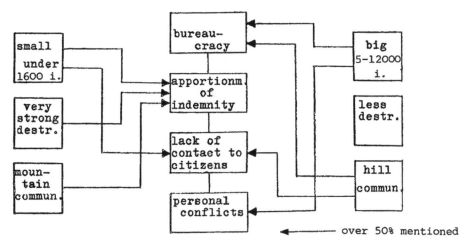

Figure 5.13 Regional differentiation of dissatisfaction.

Since a conservative majority rules in the Friuli–Venezia Giulia Region, the political rejection of it by left-leaning planners finds plain expression (Fig. 5.14). Dissatisfaction with the Region, the planners think, is even greater than with the commune.

From Stagl's studies it may be concluded, for a general theory of hazards, that a social system struck by catastrophe goes through four stages of distinct conflict potentials (Fig. 5.15). It has a certain conflict potential (phase I) when it encounters the catastrophe, while solidarity, mutual help and sacrifice minimize the conflicts (phase II). But as soon as goods are distributed in the first stage of relief (external intervention), the causes of conflict multiply (phase III) and increase during reconstruction to reach a maximum. Winners and losers in the catastrophe are pitted directly against one another.

Toward the end of the reconstruction phase the number of conflicts decreases, but there are considerably more than before the disaster. In phase IV they slowly decline, since people have become used to injustice and the

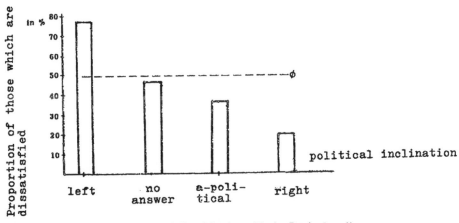

Figure 5.14 Political attitudes and dissatisfaction with the Region's policy.

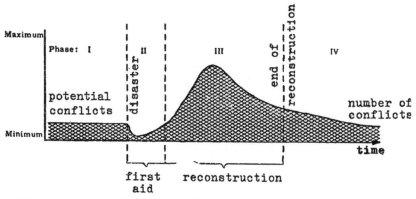

Figure 5.15 Catastrophes and changing conflicts.

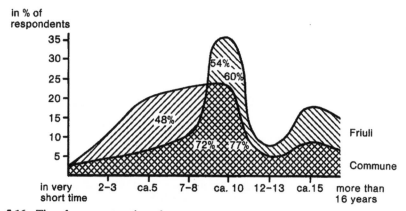

Figure 5.16 Time for reconstruction of communes and Friuli.

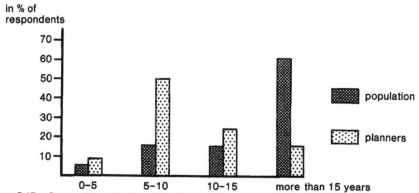

Figure 5.17 Comparison of estimations of duration of reconstruction by planners and population.

mute reproach from the old people who will live till the end of their lives in the barracks softens with time.

How long can this process last? Stagl asked the planners too about the supposed time required for reconstruction of their own communes and Friuli as a whole (Fig. 5.16).

The favoring of one's own commune – which we recognize from mental

Figure 5.18 Earthquakes are quite frequent in Italy and the shortcomings of government handling of catastrophes are notorious. On this poster, the mayors of three communes from Val Belice in Sicily invite the inhabitants of Gemona to a meeting on Sunday, March 6, 1977, at 10 a.m. in the inflatable dome (civic centre) of the Fraction of Ospedaletto. A Social Democrat, a Christian Democrat and a Communist want to compare their own experiences with that of the Gemona people and will hold a press conference about what happened to them after the earthquake they experienced in their own communities. The earthquake of Val Belice in 1968 led to a major scandal, because donations from abroad and governmental reconstruction funds disappeared in the dark channels of the Mafia and, after 10 years of waiting, people still live there in temporary crash housing.

173

map studies – shows up here, for the completion of their rebuilding is assumed to take place sooner than that of the rest of Friuli. But in the smallest communes there is particular scepticism.

Comparing studies by Stagl and Wagner[12], it is seen that planners are rather more optimistic about reconstruction than is the population (Fig. 5.17). One reason is probably the broader outlook of the planners. The inhabitants of Friuli are most aware of their own misery. The planners derive their hopefulness from the comparison with reconstruction in other disaster areas of Italy, especially from the case of Val Belice (cf. Fig. 5.18), and, maybe, from self-justification.

Ninety per cent of the planners asked believe that reconstruction in Friuli is going 'better' or even 'considerably better'. Only a few left-oriented people are sceptical even in this regard. They say that 'Nella nostra società nemenno i terremoti sono uguali per tutti' (in our society not even the earthquakes are the same for everybody).

The fact that in Friuli better reconstruction results could be expected is due, in the planners' opinion, to the mentality of the people, their greater tendency to take initiative, the absence of a mafia, the lesser wastage of materials, and decentralization of control of funds to commune level (Fig. 5.18). The talent of Friulians for greater self-rule and democratic control 'from below' may have fostered this situation. Self-rule in communes is poorly developed in Italy. Rome has authority everywhere. But the regional autonomy of Friuli established in 1963 received a further encouragement during the catastrophe. The Friulians want to take care of their political problems for themselves and therefore saw the earthquake catastrophe as their great chance to prove themselves.

5.2   Advice and recommendations for the future development of Friuli

In Section 4.3, we examined critically certain effects of external intervention. Probably the ideas that are about to follow can also be regarded as external intervention, and therefore ought to be taken with appropriate reservations by the regional authorities. Agreement with what some among them think is evident already, however, especially in regard to the issue of the continuity of settlements.

Italian experts[13] for their part favor a selective reconstruction that will simplify the network of residential communes, above all in the numerous communes in the mountains that were already half deserted before the earthquakes, '. . . which were by this time only occupied by the old people and were due to be abandoned in the short span of one or two decades . . .' .[14]

This selective reconstruction must steer between two extremes.

(a)   The former 'sprinkling-can' method of trying to restore life to a settlement that grew up in the Middle Ages on the scale of the rural land use of the time and local and long-distance trade, to which the well intentioned efforts of some of the international relief agencies (that the latter considered always correct) unconsciously gave impetus,

(b)   The proposal of 'functionalists and efficiency experts' for collecting all the reserve of people living in the prefabs into a ring of satellite settlements north of Udine surrounding a central industrial park. The area of

Osoppo was also cited as a location for such a new city of 50 000 people. The Pordenone–Codroipo–Udine–Gorizia development axis could then have been augmented by a northward trending branch from Udine to Tarvis (CML U 13).

About this, Valussi writes:

Such a solution would not only run counter to all the purposes of a culturally enlightened policy, turning the historic centers into museum ruins like Pompeii. It would also in the briefest time bring about the desertion of the mountain region and the end of agriculture in the Friulian alta pianura with the collapse of all farming activities, and the ecological desolation of that rare example of a kind of natural region that the Tagliamento moraine amphitheater represents, even if this were the most rational, economical, socially progressive solution, able suddenly to change the course of Friulian history and to alter radically the determinants of the underdevelopment of the region. Such a solution would not receive the acquiescence of the Friulian people, without whose collaboration the rebirth of Friuli is inconceivable.[15]

Valussi's attitude is supported by the findings of our study. The majority of respondents neither want to leave Friuli, nor always to live in a city, and the will to rebuild, notably of larger households and in middle age groups, is high. It may be worth noting that the younger people are rather hesitant, perhaps sceptical. Among people not owning houses who had been placed in prefabs a strong commitment to staying in the latter was found. Accordingly, the freedom of maneuver in regard to larger resettlement projects for the population will remain slight. Home- and landownership also make for strong local attachments. Willingness to rebuild is closely connected with the presence of job opportunities. It also was hardly discouraged by the loss of savings already put into reconstruction when the second earthquake came.

In Valussi's opinion, reconstruction should concentrate a simplified settlement pattern on the places that had already shown successful growth tendencies before the earthquakes and could provide a basis for a hierarchical service network that was for a long time largely lacking in Friuli.[16]

Communal reform that will abolish the decentralization of reconstruction decisions among communes that have inherited their administrative structures from a Medieval feudal and parish system only slightly simplified by the Napoleonic reform, and lacking sufficient governing power, is indispensable for this. Of the 51 communes of Pordenone Province, eight in 1975 had less than 1000 inhabitants, 15 between 1000 and 2000, and three from 2000 to 3000. Among the 137 communes of Udine Province, four had less than 500, 17 from 500 to 1000, 34 from 1000 to 2000, and 32 from 2000 to 3000 people. In these 87 communes (63 per cent of the communes of Udine Province) only 141 683 (27 per cent) of the 530 000 inhabitants lived. Among these 87 communes, a 'land redistribution' is indispensable. It is, however, greatly complicated by the fact that Regional Law 30 of June 20, 1977 and National Law 546 of August 8, 1977 expressly place the means of reconstruction in the hands of the communes.

In the literal sense too, *land redistribution* should have received a new impetus from the earthquakes. Since these brought high losses of livestock, some of the traditional combinations of enterprises (sideline industries and spare-time farming as well as regular industrial employment) were thrust by the catastrophe into an intensified process of segregation of economic activities. There is no room for keeping animals in prefabs. It must be assumed therefore that larger areas of land must have been released that could be taken up by agricultural enterprises, able because of better mechanization to cover more easily the longer distances between (now abandoned) farmstead and agricultural sites, which have resulted from concentrating the population in barrack settlements.

Dutch investigations[17] following the great flood catastrophe of 1953 have shown that such rationalization processes apparently always occur after catastrophes on a major scale, because the latter are able to bring to the fore basic questions not only concerning the 'meaning of life' but also much more modest ones about the reasons for small daily acts of decision – because of the feeling of threat and emotional disturbance of the population. As a consequence of such encounters, the Netherlands authorities were able to convince many small farmers to give up their enterprises in 1953.[18] The large public commitment of funds justifies the state's insistence on taking advantage of the reconstruction to solve such social problems as would, in the absence of catastrophe, soon have called for intervention by government.

This applies above all to the *provisions for the old age* of the many elderly people among the residents of Friuli. It would have been cruel, after the experience of 1976 and 1977, after homelessness, living in tents, evacuation and a return to prefabs, to have transplanted the no longer active population yet again. It would have been just as cruel to expect the oncoming generation to remain simply as devout preservers of an historical heritage of place in mountain communes that could not adequately be serviced.

A *social plan* for medical and old age care was supposed to insure (with such devices as 'meals on wheels') that the remaining years of older people in familiar surroundings were secure. Prefabs which became free because of deaths were, however, not to be made available and after a time would be taken away. Precise investigation of the age structure according to commune and subdivision would reveal how long this process must go on, and whether it could be made more humane by auxiliary measures (such as central old age homes in places in the middle of communes). To avoid condemning the aged into ghettos, it was also necessary to think, to a larger extent, of housing suitable for older people within new buildings, to be erected when central settlements were being rebuilt, and of making public aid depend on this (cf. the planning reported for Osoppo). There will always be a need for such housing space in Friuli, for it is part of the traditional behavior pattern for many people who leave the country to come back to retire there. In exchange for property rights in mountain districts, given up to improve the agricultural subsistence possibilities for the few remaining farmers, housing should be provided for returnees.

It will be possible to bring back a population of working age to Friuli as well and to prevent emigration only through a deliberate *industrialization* at a few selected points. For example, Tolmezzo (CML U 26) is the only commune in the mountains that can show a population increase between 1951 and 1975 (cf. Fig. 1.11), and the industrial zone of Osoppo (CML U 43), despite severe damage, has exhibited an astonishing power of regeneration (see Section 3.2, p. 92). The industrialization policy must convince the working population of Friuli that creation of new jobs does not merely mean a short-term false prosperity that will disappear when the increased demand connected with reconstruction is past, but that a branch structure is developing, the products of which will be suitable for export.

Therefore *traffic connections* with neighboring countries have to be drastically improved. The double-tracking planned for the 'Pontebbana' (Udine–Gemona–Tarvisio–Villach) and continuation of the Autobahn Venice (or Trieste)–Udine–Gemona–Villach–Tauern tunnel–Salzburg–Munich can much improve the access of Friuli to markets.

Beside this dependency on foreign connections, the development of *internal traffic* is also important. This is true for school bus systems as well as for the growing commuter traffic. It would take action space studies to show which settlement localities can still be maintained within reasonable commuting distance of the industrial centers (cf. Sec. 3.4, p. 104). The increase in automobile permits in 1974–5 in Pordenone Province (+7.7 per cent) and Udine Province (+6.0 per cent) was more than the Italian average increase of +5.3 per cent for the same period.[19] It must be taken into account as well that homelessness and evacuation, visits of foreign relatives and stays with friends and relatives both in Italy and abroad over the winter have made the population more disposed to distant contacts, enhanced their mobility and expanded their action space.

Communal reform, agrarian reform, welfare for the aged, the rearrangement of central places, industrialization, and traffic development were only the key words, intended to indicate that the public power, whether represented by commune, province, autonomous region or the central government, was called upon to make such heavy investment in these areas that the sphere of the individual initiative of citizens was necessarily reduced. The social security network and with it government tutelage always expand in times of crisis. The self-organization of public interest should be used as a counterweight. Total dependency on welfare, which was most apparent during the evacuation phase, seduced people into lethargy. Having just to wait during periods when government is indecisive makes it hard to decide for oneself later. G. Barbina[20] points in this regard to the Yugoslav reconstruction policy which succeeded – admittedly because the disaster on the Slovenian side of the border was on a smaller scale – in skipping altogether the phase of housing people in barracks. Breginj, the next commune to Taipana (CML U 46), was also fully destroyed on May 6, 1976, and there was additional damage in Kobarid (Caporetto) and the middle Isonzo Valley. The Yugoslav government concentrated all its reconstruction efforts in Breginj. Rebuilding commenced immediately and proceeded in three shifts, seven days a week. It was finished in November 1976.

A similar strategy was also under discussion in Friuli in May 1976. It was expressed in the motto 'dalle tende alle case' (out of the tents (straightaway) and into the houses), instead of barracks first. But it could be seen here that a whole new range of problems arose because of the wide fragmentation of decision making due to the areal scale of destruction in the Friulian disaster, and in view of the local settlement structure. In this instance, extrapolation from previous findings in catastrophe research is of uncertain value.[21]

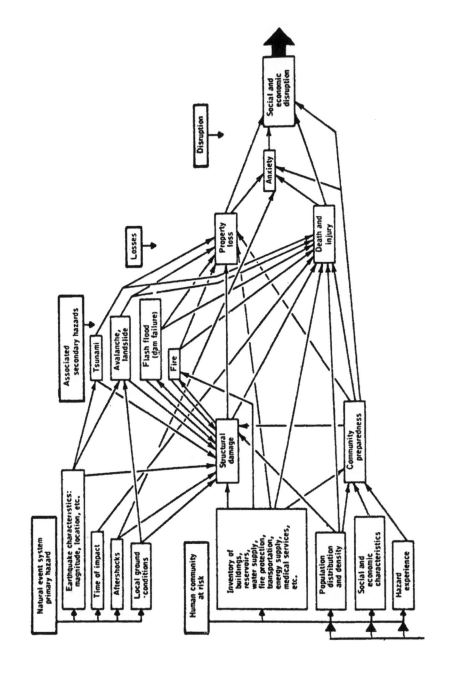

Natural event system
primary hazard

Earthquake characteristics:
magnitude, location, etc.

Time of impact

Aftershocks

Local ground
conditions

Associated
secondary hazards

Tsunami

Avalanche,
landslide

Flash flood
(dam failure)

Fire

Losses

Property
loss

Death and
injury

Disruption

Anxiety

Social and
economic
disruption

Human community
at risk

Inventory of
buildings,
reservoirs,
water supply,
fire protection,
transportation,
energy supply,
medical services,
etc.

Structural
damage

Population
distribution
and density

Social and
economic
characteristics

Hazard
experience

Community
preparedness

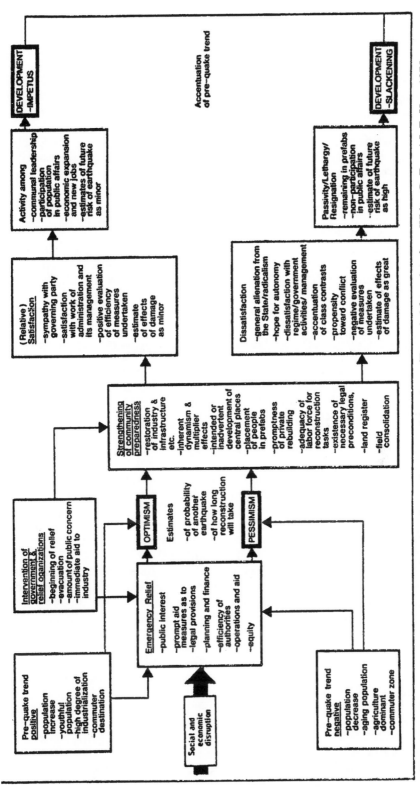

Figure 5.19 Relationships of the impact of earthquakes on human social systems. (Expanded from data from White, G. F. and J. E. Haas 1975, p. 322 – see note 24 in Chapter 2.)

The question posed in the Preface, therefore, ought to be repeated here at the end: what can the example of Friuli teach us about the relevance of the findings of recent disaster research, referred to in Chapter 2, to such occurrences in other areas (Fig. 5.19)?

Every year there take place on the Earth an average of 30 great natural catastrophes, about half of which afflict cities.[22] Reconstruction of the latter takes anything from 2 to 10 years. Certain regularities can be discerned.

(a) Persistence of the previously established spatial structure is marked.

(b) The opportunity for achieving change is great, and the intellectual input of expert opinion and planning talent is high, but fewer changes occur than one might suppose.

(c) The time required for reconstruction is a function of trends already established before the catastrophe, of the extent of damage, and of resources at hand for reconstruction.

(d) Help from outside, innovative leadership on the part of the central government, and planning measures worked out before the catastrophe can cut by half the length of time needed to restore the situation.

(e) Even though a catstrophe can be considered a 'great equalizer', under a market economy the original hierarchy of functions, persons, and power relationships asserts itself rather more sharply than ever, and it is very hard even for relief policies to change.

(f) Planners are obsessed with usually grandiose rebuilding plans, but the citizens' thinking already includes such a plan unconsciously: the plan of the city as it was before destruction. This plan competes with the new ones, especially when the latter are overly ambitious and take too long to explain in detail.

These assertions of Robert W. Kates and others apply quite broadly to events in Friuli.

But, of course, it makes a big difference if such generalizations are derived from background material in America only, or if they go back to a highly developed area where human vulnerability to disasters is governed not only by the impact of natural forces but by ethnic and class structure, political forces, tensions between local and central government and the value systems of old and very complicated historical traditions.

Our experience in Friuli appears to confirm to some extent the assertion of William I. Torry (1979)[23] that '. . . the conception, dissemination, and adaptive significance of folk beliefs and political and religious dogma deserve considerably more attention from hazard researchers than they have received to date'. The Anglophone world, especially the New World, which has supplied not only the stage for most hazard research but also the concepts and models to understand disaster impact, is a very rational one, void of the imponderabilities of people's perception of disaster. Maybe geographers working in the field of disaster research have adopted too much of the attitudes of their co-researchers from systems analysis, regional science, resource management or business administration to be able to look sensitively into the intricacies of a traditional society, not governed by the rules of cost and benefit only.

So, again, we agree with Torry when he asks questions such as, for example,

'. . . what *systems* of beliefs shape a group's understanding of hazard causality and severity, its receptivity to and absorption of novel safeguards, and its standards for choosing among contending established remedies? And in what sense do such conceptual models contribute to adaptive success and failure?'[24]
Some questions like these, we hope, have been answered by this book.
For instance, the following additional components can be added to previous lists.

(a) The persistence of the previously established spatial structure increases with the historical depth and cultural importance of a disaster area.

(b) In allocating temporary housing, care must be taken not to supply types too various in quality, because then there is a temptation to regard the better quality houses as permanent fixtures and, on the other hand, poor quality ones cause a downgrading of the people who are assigned to them and produce social tensions among people in the same plight.

(c) When there are ideological tensions in a country, political groups apply their ideas about society, most notably those having to do with property rights and construction plans derived from them, to communes where they can realize their political ideas. This produces competitive reconstruction planning and takes advantage of the 'zero-hour' laboratory situation for socio-political experimentation.

(d) The spatial situation (the 'Three countries' corner') and the peculiar cultural and linguistic status of an area, expressed in the Autonomy Statute, intensify its sense of unity and weaken the leadership rôle of innovative government power. The latter is forced in some cases to vie for the loyalty of its citizens, particularly if national minorities live in a disaster area.

(e) Long periods of seismic inactivity within a very traditional culture area allow the existing stock of buildings to go for hundreds of years without restorative tests of stress-loads. Since mostly the older and poor folk are also segregated in the oldest buildings of the early city cores, the hypothesis to the effect that a catastrophe and its consequences are extremely 'inegalitarian' events hitting the most helpless hardest is confirmed.

(f) If a disaster happens to coincide spatially with the approximate settlement area of a given ethnic group (in this case the *c.* 400 000 Friulians), it is extremely important not to forget the cultural identity of the people affected, which manifests itself in numerous features of the social-spatial situation, when dealing with the problems of geological security and architectural form. Reconstruction planning on the part of the central government should try to incorporate as much as possible of such regional individuality into its projects, which under pressure otherwise tend to foster uniformity and anonymity.

(g) Societies, as Friesema *et al.* and Wright *et al.*[25] mention in their respective books, recover rapidly after disasters. The case of Friuli shows that a catastrophe may actually inspire among the population of an area a heightened self-awareness and determination to take their fate into their own hands. The return of many emigrants from abroad, the emergence of a Friulian Regional University in Udine, increased enthusiasm for a more extensive regional autonomy, and even the appearance of an avowedly ethnic Friulian political party, attest to this effect.

On the basis of our studies we are in agreement with the injunction that 'hazard management policy stands to gain considerably from sound comparative research'.[26] We believe, moreover, that our discussion of the spatial differentiation of earthquake effects and responses is the sort of study which in the future might bring back a *geographical* view on hazard research which it might be lacking at the moment because of too high degrees of abstraction from the real world of disaster areas and their victims. Social geographers should take sides more with social psychologists, anthropologists and sociologists and less with economists and systems analysts.

Six regularities from former research have been balanced against seven points in favor of individuality from our findings. Intercultural comparison in international co-operation should invite a further deepening of hazard research. The importance of full social, cultural and historical understanding of an area to practical decisions is evident in the case of Friuli and poses a challenge to cultural, social and historical geographers to involve themselves more fully in just this sort of problem. By doing so, social geography can rescue hazard research from a dangerously naïve preoccupation with administrative technologies and procedures, and bring it back closer to the mainstream of a humanistic geography.

Notes

1 Ciborowski, A. 1967. Some aspects of town reconstruction (Warsaw and Skopje). *Impact* **17**, 31–48.
2 Douard, J. 1978. Social and administrative implications: protection, relief, rehabilitation. In *The assessment and mitigation of earthquake risk*, 285–302. Paris: UNDRO (Unesco).
3 Kates, R. W. 1976. Experiencing the environment as hazard. In *Experiencing the environment*, S. Warner, S. B. Cohen and B. Caplan (eds), 137. New York: Plenum Press.
4 Stagl, R. 1980. *Planungen und Massnahmen nach einer Katastrophe und ihre Bewertung in der Praxis. Wiederaufbaustrategien und -probleme in Friaul*. Unpubl. master's thesis, Munich.
5 *Süddeutsche Zeitung*, November 20, 1979.
6 Strassoldo, R. and B. Cattarinussi (eds) 1978. *Friuli – la prova del terremoto*, 371. Milan.
7 Kates, R. W. 1977. Reconstruction following disaster. *Natural Hazard Observer* I(4), 5.
8 Burton, I. and R. W. Kates 1972. The perception of natural hazards in resource management. In *Man, space and environment*, P. W. English and R. C. Mayfield (eds), 282–304. New York: Oxford University Press.
9 Steuer, M. 1979. *Wahrnehmung und Bewertung von Naturrisiken am Beispiel zweier ausgewählter Gemeindefraktionen im Friaul. Münchener Geographische Hefte* **43**. Regensburg: Kallmünz.
10 Ambrasey, N. N. 1976. *The Gemona di Friuli earthquake of 6 May 1976*. Restricted Technical Report FMR/CC/SC/ED/76/169, part I. Paris: Unesco.
11 Mileti, D. S., T. E. Drabek and J. E. Haas 1975. *Human systems in extreme environments: a sociological perspective*, 106. University of Colorado.
12 Wagner, U. 1979. *Wiederaufbau als Sanierungschance oder Fehlinvestition: Untersuchungen zur Mobilitätsbereitschaft in ausgewählten Abwanderungsgebieten des Friaul*, 64. Unpubl. master's thesis, Munich.
13 Valussi, G. 1977. Il Friuli di fronte alla ricostruzione. *Riv. Geog. Ital.* **LXXXIV**, 113–28.
14 Valussi, G. op. cit., p. 125.
15 Valussi, G. op. cit., p. 125.
16 Pagnini, M. P. undated. Appunti per uno studio sulla gerarchia dei centri della regione. In *Prospettive regionali*. Udine.

17 Herweijer, I. S. 1955. Het agrarisch herstel en de herverkavelingen in het rampgebied. *Tijdschrift KNAG* 72, 297–306.
18 Heslinga, M. W. 1953. Het herstel en de sanering van het rampgebied in Zuidwest-Nederland. *Tijdschrift KNAG* 70, 273–308. Heslinga shows in this paper that repair costs following the February 1953 flood (860 million hfl) would take 4 per cent of the Gross National Product of the Netherlands.
19 Regione Autonoma Friuli–Venezia-Giulia, Assessorato della Pianificazione e Bilancio 1976. *Compendio statistico*, 68. Trieste.
20 Barbina, G. 1980. Friuli centrale dopo gli sismici del 1976. *Boll. Soc. Geog. Ital.* Series X VI(10–12), 607–36.
21 Haas, J. E., R. W. Kates and M. J. Bowden (eds) 1977. *Reconstruction following disaster*. Cambridge, Mass.: MIT Press. The authors take single cities such as San Francisco, Anchorage, Managua and Rapid City as examples of where decision-making was concentrated, as opposed to Friuli with its more than 100 diverse afflicted communes.
22 Kates, R. W. 1977. Major insights: a summary and recommendation. In Haas *et al.* (eds) op. cit., pp. 261–93.
23 Torry, W. I. 1979. Hazards, hazes and holes: a critique of the environment as hazard and general reflections on disaster research. *Can. Geogs* XXIII(4).
24 Torry, W. I. op. cit., p. 379.
25 Friesema, H. P. *et al.* 1979. *Aftermath. Communities after natural disasters*. Beverly Hills: Sage.
26 Wright, J. D. *et al.* 1979. *After the clean-up. Long range effects of natural disasters*. Beverly Hills: Sage.

6 Epilogue

6.1 The facts

After the preceding chapters of this book had been written, on November 23, 1980, on a Sunday evening at 7.35 p.m. an earthquake of 6.8 R magnitude and 20 km in depth occurred in the northeastern hinterland of Naples between Avellino and Potenza in South Italy, its epicenter being 5–7 km north of Laviano. In two waves, it lasted for about 90 seconds and during the following five hours more than 200 tremors and aftershocks shook the earth. An area of about 28 000 km² was hit, with 314 communities, mainly in the Provinces of Salerno, Avellino and Potenza of the Regions of Campania and Basilicata (Fig. 6.1). About 300 000 people were made homeless; more than 3500 lost their lives.

The larger extent of the disaster and the location of the affected area distinguished Friuli from the hinterland of Naples. In Friuli, the disaster hit an area bordered by three countries with the possibility of instant aid from well informed neighbors. A large part of the Italian army was stationed right in the disaster area and was ready to help immediately. Southern Italy, on the other hand, far down at the end of the Italian 'boot' was distant from potential aid and of poor accessibility. Two thirds of the Italian army are constantly located north of the plains of the River Po, with excellent communication and good roads. The first-arriving aid groups on the Rome–Naples road were caught even before reaching the disaster area in the flood of thousands trying to escape by car from Naples, who had panicked because of the collapse of a poorly built ten-storey building and the continuing aftershocks. In spite of the rain and low temperatures, most of the people stayed outdoors until Tuesday, November 25, 1980.

So the earthquake in southern Italy also became a catastrophe of the communication and administration system. The collapse of the communication network made the disaster at first appear much smaller than it really was. Radio amateurs gave the first hints as to its real scale. Criticism of local authorities and central government grew stronger, pointing to the fact that, on a Sunday evening, Italy evidently lacks institutions which in other countries might be more responsive in such a situation.

Aid arriving from the north, including engineer troops from neighbouring countries on railway trains and in Transall airplanes, found no guidance through effective emergency offices. Nobody knew where to start. Most severely felt was the lack of heavy equipment to rescue survivors from the collapsed buildings. If it finally were to arrive in the disaster area it would have been impossible to bring it to the center of the destruction. The topography of the Apennines, with narrow mountain roads and lack of detour routes, forced the helpers to bulldoze their way first through one destroyed city centre, to come to the next, which formed a barrier again before the next to come. The helpers were too absorbed by the distress in the first community to proceed to the greater needs in the next which was more central to the epicenter. With growing peripheral location of the cities, the helplessness of

184

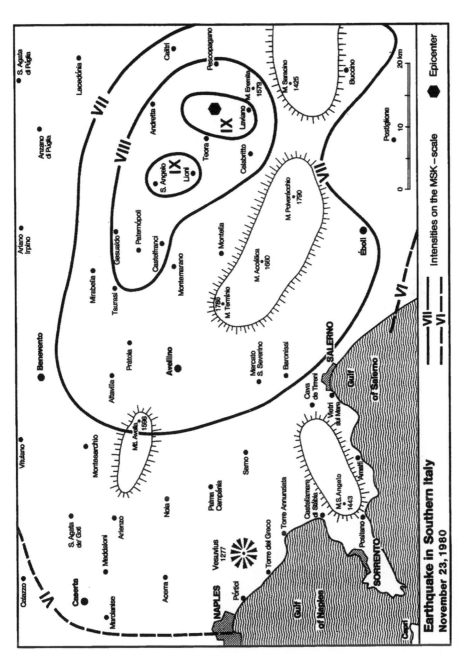

Figure 6.1 Delimitation of the disaster area in southern Italy.

the population increased too because of the age of not only people but buildings.

On Thursday November 27, 1980, after a harsh criticism of government performance by President Pertini, a Socialist, the Minister of the Interior, Rognoni, a member of the Democrazia Cristiana, responsible for the emergency operations, declared his resignation. It was rejected. On Wednesday, the Prefect of the Province of Avellino had been dismissed because of inefficiency. In spite of the fact that a law had been passed in 1970 on aid measures in case of natural disasters, by 1980 the detailed regulations were still missing. Government provided $1.2 billion for immediate relief; all over the world, especially in neighbouring European countries and in areas overseas with many Italian immigrants, collections for the victims were made.

But to bring these funds to those people really in need became a growing problem.[1] Foreign and even domestic aid organizations sometimes are denied permission to operate in a disaster area. The logistics of distributing supplies partly collapses. On the market squares of some of the destroyed cities, supply goods piled up (some of them entirely inappropriate: summer clothes for 'sunny Italy') and were soaked by the cold rain. Goods simply failed to reach other communities; it was more important to rescue those buried alive under the debris. Even four days after the earthquake, 27 people were dug out. Five days after the earthquake, 15 more emerged, and the last three were rescued 14 days after the disaster! The danger of epidemics grew because of the many unburied corpses under the debris. Because of his success in Friuli, Emergency Commissioner Zamberletti was again in charge of the emergency operations in the Naples area. He ordered railway cars to the disaster zone and appealed to the owners of trailer caravans to place them at the victims' disposal. He intended to use the strategy which had proved successful in Friuli: to commandeer hotels, condominiums and second homes on the coast to house the homeless during the winter. The number of communities to be totally evacuated was 126.

New tremors on Sunday November 30, 1980 made this evacuation to Sorrento, Amalfi, Positano, Capri or Ischia for a five-month period even more pressing. In the mountain villages, heavy snowfall set in. But only a few hundred of the homeless accepted the offer of evacuation. Distrust grew that the evacuees might be denied return to their villages in spring and that their property might be looted. Piety, too, held back many people in their ruined villages where dead relatives had not yet been exhumed, identified and buried. Container camps, ships laid up for winter in the port of Naples, and railway trains provided for the disaster victims were occupied by the homeless inhabitants of Naples, who in the chaos, pretended to be earthquake victims. There were 80 000 of them before the disaster, and the number grew, because many new buildings of very poor quality had to be declared uninhabitable and dangerous because of the numerous aftershocks.

Then the search for the guilty started. Architects and speculators, whose new buildings collapsed, and the officials of construction supervision, who closed their eyes on the poor quality, were taken to court and accused. More and more blocks of buildings similar to those that collapsed during the quake were discovered and evacuated. Forceful occupations of empty houses and apartments took place and the city government turned a blind eye. The communist mayor of Naples, Maurizio Valenzi, accused the Catholic Church of

186

not opening its many cloisters to the homeless, who instead occupied public buildings and schools. Accusations of corruption, raised even in Friuli, grew. An additional factor was the proximity of a large city like Naples, where criminal groups operated in the anonymity of the chaos, and not only looted ruins but stole supply goods and bought cattle from distressed and homeless farmers for extremely low prices. The 'Camorra', a counterpart to Sicily's Mafia, took its share of the profits that are possible in the aftermath of a catastrophe. Summary courts were established to try looters and profiteers. Public concern grew for the more than 1000 orphans, some of whom had been taken away in private cars. Was it eagerness to help, or kidnapping? In a situation of incipient anarchy, the Camorra offered itself as a kind of 'sottogoverno' or 'government below the government'[2] and was sometimes more effective than the authorities. The general situation may be illustrated by the case of a technically well equipped relief team from the city of Siena. Threatened with force, the team was refused permission to begin rescue operations by the administration of a destroyed village. The goodwill team had to leave because the clearing of debris had already been contracted for payment with a demolition firm from Naples.[3]

The weak performance of the Italian government becomes more understandable if we acknowledge the fact that the disaster area had long since lived on subsidies, most of which had been 'siphoned off' in the regional capitals of Naples and Potenza.[4] Should this part of southern Italy be reconstructed at all in its medieval settlement pattern? The acropolis-like location of most of the old, densely built cities on the summits of steep hills, which made sense in times of war, was a dangerous anachronism in itself. But reconstruction on the same site is vital to a population deeply rooted in tradition, while the more enterprising members of the local society had long since left the region. The 'Mezzogiorno' or Italy's southernmost region was industrialized only in its most accessible parts, mainly in the coastal areas rather than the hinterland. This can be taken from the fact that of the supposed $16 billion total damage, only less than half a billion was lost by industry. Economists, among them the Minister of Budget, La Malfa, therefore proposed to 'make one clean sweep' of the whole region. But this implied a lengthy planning period whereby the imminent needs of the population could not be met. First of all, corruption, nepotism and misconduct in office should be rooted out, to gain the trust of potential investors in the development of southern Italy. As long as private business of northern Italy was unwilling to put investment into the chaos, the European Community, World Bank and international financial groups could not be reproached for being reluctant.

This chaos was aggravated by the fact that besides the Camorra, the terrorist 'Red Brigades' also interfered. On May 8, 1981, the Christian Democrat, Ciro Cirillo, who had been responsible for the reconstruction and who was taken hostage 11 days earlier, gave a first sign of life through letters, in which he demanded that vacant houses in Naples and the disaster area should be confiscated and given to the homeless, until the state could provide sufficient housing for the victims. The Italian Parliament, in a session that started on December 4, 1980, discussed the responsibility for the social catastrophe that had followed the natural disaster so dramatically.

The credibility of a government, of which some members were compromised by scandals, was questioned. The Communist Party (which had long been pressing towards the 'historical compromise' of partaking in a government led by the Democrazia Cristiana) for the first time demanded a new government and offered itself as 'savior from the national plight'.

This was what happened between November 23, 1980, and May 1981 within six months. What are the lessons for a comparison?

6.2   The lessons

In but one country we have observed that there are markedly different responses to the same hazard. The reasons for these differences are complex, but this analysis provides some insights. There are four major factors: (a) the physical characteristics; (b) administrative and organizational reasons; (c) social and economic reasons; (d) the characteristics of the inhabitants, their attitudes and beliefs.

(a)   The most striking difference is the size of the area affected and its location within Italy. The area was six times greater than in Friuli and very remote. The higher losses in southern Italy may be attributed to the fact that most settlements were urban with houses of four or five storeys. Many newer buildings collapsed because of their poor quality. The sites of most settlements on the summits of hills led to snowball effects when buildings collapsed. Table 6.1 provides some physical facts for a characterization.

(b)   Looking for indicators of the administrative power of the regions under comparison, it is striking that the first decree of a three-category delimitation of the disaster area was issued in Friuli 12 days after the earthquake. In southern Italy, it took three months and led to only two categories. The deadlines indicated in Table 1.3 were not met. This led to tensions between the aid coming from outside and the ineffective local emergency authorities, regarding priorities of such help. The size of the disaster area led to the arrangement that, from the very

Table 6.1   Comparison of Friuli 1976 and southern Italy 1980.

Facts	Friuli	southern Italy
date	May 6, 1976	November 23, 1980
time	9.00 p.m.	7.35 p.m.
magnitude	6.4 R	6.8 R
depth	c. 5 km	c. 20 km
duration	55 s	90 s
epicenter	2 km west of Venzone	between Lioni/Teora
affected area	4800 km^2	28 000 km^2
number of communities	119	314
deaths	1000	3500
homeless	100 000	300 000
damage ($ billion)	4.45	15.9 (estimate)
buildings	1.72	5.4–8.1
public infrastructure	0.90	4.0–4.5
industry	1.32	0.4
handicraft	0.02	0.2

beginning, the independent responsibility of the community had priority over all centralized measures. But these local activities were of course influenced by the capabilities and good connections of the respective mayors, as well as by their shortcomings and corruptibility.

(c) What was possible under the more intimate, homogeneous conditions of Friuli failed in the environment of a city of more than a million inhabitants, a city characterized by extreme contrasts of rich and poor, social class and stratification. Without an earthquake, Naples already had approximately 80 000 homeless within its municipal boundaries. Plans for evacuation soon turned into the occupation of private property, with the state as an accomplice during and after the fact. Quantity makes for a new quality of problems. The homeless of Naples alone (10 per cent of the population) equalled the whole afflicted population of Friuli.

(d) Of course there are many exceptions to the rule, but experiences of aid organizations shed light on the differences of mentality which are rarely dealt with in studies on disasters. While, in Friuli, solidarity between victims and helpers was soon established, the afflicted population in southern Italy was distrustful of both state and international aid. A recurring comment was 'What good will come out of Rome?'. Fatalism and apathy were much more common than in Friuli. Loyalty hardly reached beyond family ties. Envy and jealousy inhibited mutual decisions and actions. Some mayors are said to have rejected the offer of prefabricated houses because they foresaw ensuing conflicts in distributing them among their citizens.

Another example was the impossibility of enforcing an effective evacuation, which was a success in Friuli but failed in southern Italy because of the pervasive distrust of the victims and partly because of the resistance of the owners of second homes.

6.3 Conclusions

Evidently, variables such as 'civic maturity', 'neighborly solidarity' and 'administrative integrity' (difficult to operationalize as they are) should be given more important weighting in all models of disaster management, if only as unknown quantities. Even events in the same nation, and under the administration of the same emergency commissioner, are not guaranteed to take the same course.

Notes

1 Widman, C. 1980. Die Stunde der Samariter und Schakale. *Süddeutsche Zeitung*, December 5.
2 Widmann, C. 1980. Obdachlose vor dem Zugriff der Camorra. *Süddeutsche Zeitung*, December 28.
3 *Handelsblatt*, March 18, 1981.
4 Gröteke, F. 1981. Italiens schwer geprüfter Süden. *Süddeutsche Zeitung*, March 21.

Appendix

A.1 Commune master list (CML)

No.	No. on map	Commune	No.	No. on map	Commune
Udine Province (U)					
1	120	Aiello del Friuli	49	96	Lestizza
2	34	Amaro	50	137	Lignano Sabbiadoro
3	22	Ampezzo	51	9	Ligosullo
4	136	Aquileia	52	41	Lusevera
5	17	Arta Terme	53	51	Magnano in Riviera
6	44	Artegna	54	48	Majano
7	59	Attimis	55	12	Malborghetto Valbruna
8	119	Bagnaria Arsa	56	93	Manzano
9	90	Basiliano	57	135	Marano Lagunare
10	101	Bertiolo	58	78	Martignacco
11	104	Bicinicco	59	84	Mereto di Tomba
12	37	Bordano	60	27	Moggio Udinese
13	49	Buia	61	81	Moimacco
14	92	Buttrio	62	45	Montenars
15	100	Camino al Tagliamento	63	103	Mortegliano
16	91	Campoformido	64	65	Moruzzo
17	121	Campolongo al Torre	65	124	Muzzana del Turgnano
18	131	Carlino	66	53	Nimis
19	57	Cassacco	67	43	Osoppo
20	110	Castions di Strada	68	15	Ovaro
21	33	Cavazzo Carnico	69	66	Pagnacco
22	6	Cercivento	70	123	Palazzolo dello Stella
23	127	Cervignano del Friuli	71	112	Palmanova
24	107	Chiopris-Viscone	72	7	Paluzza
25	28	Chiusaforte	73	85	Pasian di Prato
26	82	Cividale del Friuli	74	10	Paularo
27	95	Codroipo	75	98	Pavia di Udine
28	56	Colloredo di Monte Albano	76	117	Pocenta
29	4	Comeglians	77	11	Pontebba
30	94	Corno di Rosazzo	78	118	Porpetto
31	76	Coseano	79	69	Povoletto
32	75	Dignano	80	97	Pozzuolo del Friuli
33	19	Dogna	81	86	Pradamano
34	63	Drenchia	82	3	Prato Carnico
35	24	Enemonzo	83	130	Precenicco
36	70	Faedis	84	87	Premariacco
37	64	Fagagna	85	30	Preone
38	134	Fiumicello	86	8	Prepotto
39	83	Flaibano	87	60	Pulfero
40	42	Forgaria nel Friuli	88	47	Ragogna
41	1	Forni Avoltri	89	5	Ravascletto
42	20	Forni di Sopra	90	23	Raveo
43	21	Forni di Sotto	91	68	Reana del Roiale
44	40	Gemona del Friuli	92	80	Remanzacco
45	111	Gonars	93	36	Resia
46	62	Grimacco	94	35	Resiutta
47	129	Latisana	95	2	Rigolato
48	25	Lauco	96	55	Rive d'Arcano

No.	No. on map	Commune	No.	No. on map	Commune
97	109	Rivignano	118	13	Tarvisio
98	122	Ronchis	119	67	Tavagnacco
99	128	Ruda	120	116	Teor
100	54	San Daniele del Friuli	121	132	Terzo d'Aquileia
101	125	San Giorgio di Nogaro	122	26	Tolmezzo
102	99	San Giovanni al Natisone	123	71	Torreano
103	73	San Leonardo	124	126	Torviscosa
104	72	San Pietro al Natisone	125	39	Trasaghis
105	105	Santa Maria la Longa	126	8	Treppo Carnico
106	114	San Vito al Torre	127	50	Treppo Grande
107	77	San Vito di Fagagna	128	58	Tricesimo
108	14	Sauris	129	106	Trivignano Udinese
109	61	Savogna	130	79	Udine
110	89	Sedegliano	131	108	Varmo
111	29	Socchieve	132	38	Venzone
112	74	Stregna	133	32	Verzegnis
113	16	Sutrio	134	31	Villa Santina
114	46	Taipana	135	133	Villa Viventina
115	102	Talmassons	136	113	Visco
116	115	Tapoglicno	137	18	Zuglio
117	52	Tarcento			

Pordenone Province (P)

No.	No. on map	Commune	No.	No. on map	Commune
1	14	Andreis	27	50	Morsano al Tagliamento
2	19	Arba	28	46	Pasiano di Pordenone
3	35	Arzene	29	12	Pinzano al Tagliamento
4	22	Aviano	30	25	Polcenigo
5	45	Azzano Dècimo	31	38	Porcia
6	13	Barcis	32	39	Pordenone
7	41	Brugnera	33	42	Prata di Pordenone
8	26	Budoia	34	51	Pravisdomini
9	30	Càneva	35	32	Roveredo in Piano
10	40	Casarsa della Delizia	36	37	Sacile
11	11	Castelnovo del Friuli	37	28	San Giorgio della Richinvelda
12	18	Cavasso Nuovo	38	29	San Martino al Tagliamento
13	47	Chions	39	27	San Quirino
14	1	Cimolais	40	44	San Vito al Tagliamento
15	2	Claut	41	20	Sequals
16	10	Clauzetto	42	48	Sesto al Reghena
17	33	Cordenons	43	24	Spilimbergo
18	49	Cordovado	44	3	Tramonti di Sopra
19	6	Erto e Casso	45	4	Tramonti di Sotto
20	17	Fanna	46	9	Travesio
21	43	Fiume Veneto	47	21	Vajont
22	31	Fontananfredda	48	36	Valvasone
23	7	Frisanco	49	5	Vito d'Asio
24	16	Maniago	50	23	Vivaro
25	8	Meduno	51	34	Zoppola
26	15	Montereale Valcellina			

A.2 The questionnaire

1. In which community do you live? ⬜⬜⬜⬜ 1/4

.....................⁄..............

2. How old are you? years ⬜⬜⬜ 5/7
3. Sex male ₁⬜ female ₂⬜ 8
4. How many relatives, apart from yourself, live here with you? ⬜⬜ 9/10
5. Did you lose any near relatives as a result of the earthquake? yes ₁⬜ no ₂⬜ 11
6. Did you lose your job as a result of the earthquake? yes ₁⬜ no ₂⬜ 12
7. Did you own your house before the earthquake? yes ₁⬜ no ₂⬜ 13
8. Did you own a non-agricultural business before? yes ₁⬜ no ₂⬜ 14
9. Do you own land in your commune of residence? yes ₁⬜ no ₂⬜ 15
10. Is the main wage earner of your family at present in employment?
    yes, in Friuli ₁⬜ 16
    yes, other regions of Italy ₂⬜
    yes, in a foreign country ₃⬜
    no ₄⬜
11. Are you in the same place of work as before the earthquake? yes ₁⬜ no ₂⬜ 17
12. Is any member of your family receiving a pension? yes ₁⬜ no ₂⬜ 18
13. Which earthquake destroyed your previous apartment or house?
    only the May earthquake ₁⬜ 19
    only the September earthquake ₂⬜
    both earthquakes ₃⬜
    none of these ₄⬜
14. Has some member of your family already taken the initiative for the reconstruction or repair of your old dwelling?
    yes, after the May earthquake ₁⬜ 20
    yes, after the September earthquake ₂⬜
    yes, after both earthquakes ₃⬜
    no ₄⬜
15. What initiative do you intend to take in the near future to improve your present situation?
    – begin immediately the reconstruction of your old dwelling ₁⬜ 21
    – move immediately into another part of Friuli ₂⬜
    – move permanently into another area outside of Friuli ₃⬜
    – move away temporarily from Friuli and return later ₄⬜
    – remain in the emergency building ₅⬜
16. In what type of house would you prefer to live in the future
    – in a wooden house resistant to earthquakes? ₁⬜ 22
    – in a cement/brick dwelling resistant to earthquakes? ₂⬜
17. Could you imagine yourself living for the rest of your life in a town like Udine or Trieste? yes ₁⬜ no ₂⬜ 23

⬜ 24

192

Further reading

Atteslander, P. 1974. *Methoden der empirischen Sozialforschung*. Berlin: Springer.

Ayre, R. S. 1975. *Earthquake and tsunami hazards in the US. A research assessment*. Monograph NSF–RA–E–75–005, University of Colorado Program on Technology, Environment and Man.

Baker, E. J. 1975. *Land use management and regulations in hazardous areas. A research assessment*. Monograph NSF–RA–E–75–008. University of Colorado Program on Technology, Environment and Man.

Baker, E. J. 1976. *Toward an evaluation of policy alternatives governing hazard-zone land uses*. Natural Hazard Research Working Paper 28, Toronto.

Barker, M. and I. Burton 1969. *Differential response to stress in natural and social environments: an application of a modified Rosenzweig picture-frustration test*. Natural Hazard Research Working Paper 5, Toronto.

Bonifacio, G. 1972. *Aspetti demografico-economici della regione Friuli–Venezia Giulia*. In IFRES (1972), op. cit., S. 88–103.

Building and Safety Division, County of Los Angeles 1975. *Building laws*. Los Angeles, USA.

Burton, I. 1968. *The human ecology of extreme geophysical events*. Natural Hazard Research Working Paper 1, Toronto.

Caporiacco, G. di 1967. *Storia e statistica dell'emigrazione dal Friuli e dalla Carnia*, vol. 1. Udine.

Cavazzani, A. 1981. *Social and institutional impact of the 1980 earthquake in southern Italy. Problems and prospects of civil protection*. Paper presented to the third international conference on The Social and Economic Aspects of Earthquakes and Planning to Mitigate their Impact, Bled, Yugoslavia, July.

City of Rialto Planning Department 1975. *Seismic and public safety element: preliminary study*. Rialto, August.

Cochran, A. 1972. *A selected, annotated bibliography on natural hazards*. Natural Hazard Research Working Paper 22, Toronto.

Cochrane, H. C. 1975. *Natural hazards and their distributive effects*. Monograph NSF-RA-E-75-003. University of Colorado Program on Technology, Environment and Man.

Cochrane, H. C., J. E. Haas, M. J. Bowden and R. W. Kates 1972. *Social science perspectives on the coming San Francisco earthquake: economic impact, prediction and reconstruction*. Natural Hazard Research Working Paper 25, Toronto.

Comunita Collinare del Friuli 1976. *Prime ipotesi del Friuli*. Majano.

Cremonesi, A. 1977. *Storia di terremoti in Friuli*. Udine.

Department of City Planning, City of Los Angeles 1975. *Seismic safety plan*. Los Angeles, USA, September 10.

Disaster Research Center (DRC) 1977. *Publications, Part A*. Columbus, Ohio, USA, July.

Dworkin, J. 1972. *Global trends in natural disaster*. Natural Hazard Research Working Paper 26, Toronto.

Earthquake Engineering Research Institute 1975. *Learning from earthquakes. Social science field guide*, vol. 4. Working draft.

Ericksen, N. J. 1975. *Scenario methodology in natural hazards research*. Monograph NSF-RA-E-75-010. University of Colorado Program on Technology, Environment and Man.

Federal Disaster Assistance Administration and California Office of Emergency Services 1975. *Los Angeles Orange Counties earthquake planning project: overview of estimated earthquake damages and casualties*. NOAA report synthesis, May.

Gazaerro, M. L. 1977. Il rapporto uomo-calamità naturali, in riferimento a una recente indagine internazionale. *Riv. Geog. Ital.* **LXXXIV** (1), 139–53.

Geipel, R. 1977. Erdbebenrisiko in Kalifornien – Einfluss auf Stadtplanung und Wirkung von Voraussagen auf die regionale Wirtschaft. *Geog. Rundschau* **29**(1), 2–9.

Geipel, R. 1980a. Aspetti geografici della percezione ambientale. In *Ricerca geografica e percezione dell'ambiente*, R. Geipel *et al.* (eds). Milan.

Geipel, R. 1980b. La percezione del rischio di terremoto. In Geipel (1980), op. cit.

Golant, S. 1969. *Human behavior before the disaster. A selected annotated bibliography*. Natural Hazard Research Working Paper 9, Toronto.

Golant, S. and I. Burton 1969. *Avoidance-response to the risk environment*. Natural Hazard Research Working Paper 6, Toronto.

Gottschalt, F. 1979. *Folgewirkungen einer Katastrophe und ihre Bewertung durch industrielle Entscheidungsträger*. Unpubl. master's thesis, Munich.

Grether, D. M. 1976. *Social and economic implications of earthquake predictions*. Interim report no. 2, California Institute of Social Sciences, Pasadena, Research Proposal no. 69318 to Jet Propulsion Laboratory, June 1.

Haas, J. E. 1976. *The consequences of large-scale evacuation following disaster: Darwin, Australia cyclone*. Natural Hazard Research Working Paper 27, Toronto.

Haas, J. E. and D. S. Mileti 1976a. *Summary of selected findings from research on the socioeconomic and political consequences of earthquake-prediction*. First draft.

Haas, J. E. and D. S. Mileti 1976b. *Consequences of eathquake-prediction*. Paper presented at Australian Academy of Science Symposium on Natural Hazards in Australia, Section 3: Community and Individual Responses to Natural Hazards, Canberra, May 26–9.

Haas, J. E. and D. S. Mileti n.d. *Socioeconomic impact of earthquake-prediction on government, business and community*. University of Colorado.

Hard, G. 1969. Die Diffusion der Idee der Landschaft'. Präliminarien zu einer Geschichte der Landschaftgeographie. *Erkunde* 23, 249–363.

Hewitt, K. 1969a. *Probabilistic approaches to discrete natural events: a review and theoretical discussion*. Natural Hazard Research Working Paper 8, Toronto.

Hewitt, K. 1969b. *A pilot study of global natural disasters of the past twenty years*. Natural Hazard Research Working Paper 11, Toronto.

IFRES 1972. *La realtá sociale di una regione in fase di sviluppo*. Atti 1. Convegno IFRES, Udine.

Istituto per l'Encicopedia del Friuli–Venezia Giulia 1971–4. *Enciclopedia monografica del Friuli–Venezia Giulia*. Part 1, *Il paese* (2 vols), Udine 1971; Part 2, *La vita economica*, vol. 1 Udine 1972, vol. 2 Udine 1974.

Kates, R. W. 1970. *Natural hazard in human ecological perspective: hypotheses and models*. Natural Hazard Research Working Paper 14, Toronto.

Kates, R. W. 1973. *Human impact of the Managua earthquake disaster*. Natural Hazard Research Working Paper 23, Toronto.

Keeble, D. 1968. Models of economic development. In *Socio-economic models in geography*, R. J. Chorley and P. Haggett (eds). London: Methuen.

Lee, T. 1973. Psychology and living space. In *Image and environment*, Downs and Stea (eds). Chicago: Aldine.

Los Angeles Earthquake Commission 1971. *Report of the Los Angeles Earthquake Commission – San Fernando earthquake, February 9, 1971*. Los Angeles, November.

Mileti, D. S. 1975. *Human systems in extreme environment*. Monograph 21, University of Colorado Program on Technology, Environment and Man.

Mitchell, W. A. 1976. *The Lice earthquake in southeastern Turkey: a geography of the disaster*. United States Air Force Academy, December.

Mukerjee, T. 1971. *Economic analysis of natural hazards: a preliminary study of adjustments to earthquakes and their costs*. Natural Hazard Research Working Paper 17, Toronto.

Natural Hazard Research 1970. *Suggestions for comparative field observations on natural hazards*, revised edn. Natural Hazard Research Working Paper 16, Toronto, November.

Nimis, G. P. 1976. *Gemona del Friuli. Appunti per una ricostruzione*. Udine.

O'Keefe, P., K. Westgate and B. Wisner 1976. Taking the naturalness out of disasters. *Nature* **260**.

O'Riordan, T. 1971. *The New Zealand Earthquake and War Damage Commission – a study of a national natural hazard insurance scheme*. Natural Hazard Research Working Paper 20, Toronto.

Pagani, B. M. 1968. *L'emigrazione friulana dalla metà del secolo XIX al 1940*. Udine.

Pagnini, A. M. P. 1975. *Sottosviluppo e terremoto. La valle del Belice*. Palermo.

Porter, P. W. 1978. *The ins and outs of environmental hazards*. Working Paper 3, University of Minnesota.

Progress report: collaborative research on natural hazards 1973. Natural Hazard Research Working Paper 31, Toronto.

Pro loco e del bollettino Parrochiale di Osoppo: Osoppo 76. Osoppo, August 1976.

Prost, B., 1973. *Le Frioul, région d'affrontements*. Geneva.

Regione Autonoma Friuli–Venezia Giulia 1974–6. *Compendio statistico: Friuli–Venezia Giulia*. Issued annually.

Regione Autonoma Friuli–Venezia Giulia, Assessorato de Lavori Pubblici 1976. *Esempi di intervento per la riparazione e il rafforzamento antisismico di edifici di abitazione*. Trieste.

Regione Autonoma Friuli–Venezia Giulia, Segreteria Generale Straordinaria n.d. *Relazione di sintesi sisma del 6 Maggio*. Prof. arch. L. Di Sopra, Udine.

Rhode-Jüchtern, T. 1975. Geographie und Planung. *Marburger Geog. Schrift.* **65.**

Robinson, R. 1974. Toward earthquake prediction. *Foresight*, November/December. Washington, DC: The Defense Civil Preparedness Agency, The Pentagon.

Ronza, R. 1976. *Friuli – dalla tende al deserto!?* Milan.

Russell, C. S. 1969. *Losses from natural hazards*. Natural Hazard Research Working Paper 10, Toronto.

San Bernardino County 1974. *Emergency plan*. San Bernardino.

San Francisco Department of City Planning 1974. *Community safety plan for the comprehensive plan of San Francisco. A proposal for citizen review.* July.

Schiff, M. R. 1970. *Some theoretical aspects of attitudes and perception.* Natural Hazard Research Working Paper 15, Toronto.

Sopra, L. Di 1967. *La struttura urbanistica friulana.* Udine.

Sopra, L. Di 1977. La baraccopoli più grande d'Europa. *Ricostruire* no. 1, April.

Stagl, R. 1981. Terromoto e ricostruzione secondo gli uffici tecnici dei 45 comuni disastrati. *Ricostruire*, anno quinto, no. 15, 8–9.

Steinbrugge, K. V. 1968. *Earthquake hazards in the San Francisco Bay area: a continuing problem in public policy.* Institute of Governmental Studies, University of California, Berkeley.

Turner, R. H. 1976. *Mobilizing the masses.* Draft, July.

United States Department of Commerce 1973. *A study of earthquake losses in the Los Angeles, California area.* Report prepared for the Federal Disaster Assistance Administration, Washington, DC.

United States Department of the Interior 1976. *Earthquake prediction – opportunity to avert disaster.* Conference on Earthquake Warning and Response, San Francisco, California, November 7. USGS Circular 729.

Valussi, G. 1961. *L'emigrazione in Valcellina (Friuli).* Firenze.

Valussi, G. 1967. *Friuli–Venezia Giulia. Collana di bibliografie geografici di regione.* Naples.

Valussi, G. 1971a. *Friuli–Venezia Giulia.* Turin.

Valussi, G. 1971b. Friuli–Venezia Giulia: ambiente geografico generale. In *Enciclopedia monografica del Friuli–Venezia Giulia*, Part 1: *Il paese*, vol. 1, 19–58. Udine.

Valussi, G. 1971c. La populazione del Friuli–Venezia Giulia. In *Enciclopedia monografica del Friuli–Venezia Giulia*, Part 1: *Il paese*, vol. 2, 759–805. Udine.

Valussi, G. 1971d. Il fenomeno migratorio in Friuli fra i processi di deruralizzazione e industrializzazione. In *La realtá sociale di una regione in fase di sviluppo*, 104–26. Atti 1. Convegno IFRES, Udine.

Vaughan, C. K. 1971. *Notes on insurance against loss from natural hazards.* Natural Hazard Research Working Paper 21, Toronto.

White, G. F. 1973. Natural hazards research. In *Directions in geography*, R. J. Chorley (ed.), 193–216. London: Methuen.

Zorzi, P. 1976. I resultati del Convegno del CISM sugli eventi sismici del Friuli–Venezia Giulia. In *Rassegna tecnici del Friuli–Venezia Giulia XXVII.* Udine.

Index

Page numbers for tables are shown in italics.

197

199

Printed and bound by CPI Group (UK) Ltd, Croydon, CR0 4YY

17/10/2024

01775656-0011